GEOGRAPHY
ITS HISTORY AND CONCEPTS

GEOGRAPHY
ITS HISTORY AND CONCEPTS
A Student's Guide

Arild Holt-Jensen

Senior Lecturer in Geography
University of Bergen

English adaptation and translation by Brian Fullerton
Senior Lecturer in Geography
University of Newcastle-upon-Tyne

Harper & Row, Publishers
London

Cambridge
Hagerstown
Philadelphia
New York

San Francisco
Mexico City
Sao Paulo
Sydney

First published in the UK 1981
Reprinted 1982, 1984, 1985

Harper & Row Ltd
28 Tavistock Street
London WC2E 7PN

British Library Cataloguing in Publication Data
Holt-Jensen, Arild
    Geography.
    1. Geography – History
    2. Geography – Methodology
    I. Title
    II. Geografiens innhold og methoder. *English*
    910'.01   G80

    ISBN 06-318186-X
    ISBN 06-318187-8 Pbk

Typeset by Inforum Ltd, Portsmouth
Printed and bound in Great Britain by
The Bath Press, Avon

To my teacher in geography
Professor Fridtjov Isachsen
1906–79

ACKNOWLEDGEMENTS

The author and the publishers would like to thank the following for permission to reproduce copyright material:

Edward Arnold for Figures 4 and 12, from Harvey, D. *Explanation in Geography,* 1969 pp 34 and 454; and Figure 5, from Haggett, P., *Locational Analysis in Human Geography,* 1965, p.14.

Methuen & Company Ltd. for Figure 13, from Chorley, R.J., *Directions in Geography* 1973, p. 38 and Figure 9, from Haggett, P., *Models in Geography* 1967, p 553.

The Geographical Association for Figure 10, from Kirk, W., *Geography* Vol. 48 p. 364, 1973.

Pergamon Press for Figure 6, from Forer, P., *Progress in Human Geography,* Vol. 2 p. 247 1978.

CONTENTS

PREFACE

Recent years have seen a more active discussion of the philosophy and methodology of geography, and there has been a corresponding increase in the number of books and papers in this field. Much of this literature is written by specialists for specialists and assumes a knowledge of the history of geography and of the development of philosophical and methodological concepts which a new student will not yet have. This book is intended to serve as an introduction to the way geographers have thought and now think, and to the methods they use. It also seeks to compare developments in geographical thought with contemporary developments in other disciplines.

The four main chapters of the book cover: the history of ideas in geography from antiquity to the present day; the 'quantitative revolution' in relation to the concept of 'paradigms' in science; the new 'critical revolution' with its emphasis on the influence of values on research; the search for geographical synthesis as it has developed from simple description to systems analysis. The book restricts itself to basic explanations; it does not cover techniques (in cartography, fieldwork or statistical analysis) which may be needed at a specific stage of a research project for which there are already many good manuals.

It is always difficult to establish a satisfactory balance between an easily comprehensive account and a rigorous use of concepts in an introductory text. The problem is made worse in so far as authors who are cited in the bibliography have worked with slightly different definitions of several concepts which are central to geography. The reader may therefore still detect some inconsistencies. To help the student to monitor differences in the use of concepts among geographical thinkers and also for quick reference, there is a subject index (p166) which lists the pages on which concepts are defined. These concepts are italicised in the text, when they are introduced. An author and personality index provides brief biographical details of writers on geography mentioned in the text.

An introductory work which is to be readable and also comprehensible to students must similarly simplify the story of the interactions between scholars. The sheer number of scientists in the modern world and the multiplicity of contacts between them through books, conferences and academic visits makes it impossible to appreciate all the influences of one scholar upon another or to explain in detail why different trends in research emerged. I have been obliged to choose what I consider to be the major trends and influences, mentioning only a small number of the scientists who have taken part in the development of geography and ignoring a number of personal contacts which may well have had a significant impact on the events described in the text. The reader should therefore be warned against the temptation to simplify the account even further in his own mind and so add to the many myths already in circulation about who or what influenced whom, which the persons concerned would like to modify or deny.

This book has developed through a number of stages. When I began to teach an elementary course on the history and philosophy of geography in Bergen in 1970, I could find no satisfactory text for students in either the English or Scandinavian languages. I therefore started to prepare a compendium on the basis of my lectures. Comments and corrections from colleagues and students followed by more systematic reading on my own part led to the publication of *Geografiens innhold og metoder* (Norwegian University Press, 1976). This book had a promising reception among Scandinavian students and also attracted a number of further comments and suggestions for improvement and alteration from fellow geographers. Starting from the Norwegian text, Brian Fullerton translated and supervised an English edition. The preliminary translation encouraged me to revise the book substantially and to enlarge it. This book is therefore the result of a joint project, but responsibility for any defects or false conclusions rests wholly with the author.

I would also like to acknowledge the debts I owe, in particular to Fridtjov Isachsen, Jens Christian Hansen, Ove Biilmann, Aadel Brun Tschudi, Torsten Hägerstrand, Magne Helvig, Jan Lundquist, Axel Sömme, Hans Skjervheim, Michael Morgan and Leslie Hesse for good advice.

   Undoubtedly, though, my biggest debt is to my wife, Elisabeth, and our two children, for their patience and forebearance with a husband and father absorbed in his geographical world of ideas both day and night.

Arild Holt-Jensen
Bergen
November 1980

CHAPTER ONE

WHAT IS GEOGRAPHY?

Most people have very vague notions about the content of scientific geography. School geography has left many with bad memories of learning the names of rivers and towns by rote. It is still common to meet people who think that geographers must have to learn a mass of facts, must know the population of towns all over the world and can name and locate all the new states in Africa. This idea of geography as an encyclopaedic knowledge of places is illustrated when a newspaper rings up its local department of geography to find out how many towns there are in the world called after Newcastle, or when readers write in to settle bets as to which is the world's longest river.

People also have an idea that geography has something to do with maps. Less cynically than Swift:

So geographers, in Afric-maps
With savage-pictures fill their gaps
And o'er unhabitable downs
Place elephants for want of towns.

Geographers are thought to be people who know how to draw maps and are somehow associated with the Ordnance Survey or the US Coast and Geodetic Survey.

A third view is that geographers write travel descriptions — a reasonable view for anyone who reads reviews of the year's books and sees that most of those listed under 'geography' are accounts of eventful expeditions to the head-hunters of the Amazon or to similar peoples and places.

Each of these three popular opinions as to what geography is has some truth in it. The names and locations of towns are facts for geographers of the same order as dates are facts for historians. They are the basic building blocks of the subject, but they are not the subject itself. The map, which represents a collection of such data, is the geographer's most useful specific resource. Different types of thematic maps are also important means of expression in geographical research, along with tables, diagrams and written accounts. Observations recorded during travel and fieldwork still provide essential data for geographers. A cultivation of the powers of observation is therefore an important objective in the education of a geographer. It is of primary importance to learn how to 'see geographically' — to observe and interpret a natural or cultural landscape without having to have local knowledge of it. Geographical training develops a distinct method of observation aimed at the understanding of the form of culture and landscape in quite a different way from that of the ordinary tourist. The lack of ability to observe in this way distinguishes most travel writers from geographers. Admittedly, many people gained their first interest in geography through reading travel stories at school, but those who later trained as geographers eventually came to disdain this type of light reading. The absence of relevant observation becomes irritating and ensures that not much geography can be learned from most travel books.

Before the present century, voyages of discovery and the mapping of formerly unknown lands were, however, closely associated with geography, Wayne K Davies (1972, p11), for instance, maintains that geography enjoyed its strongest relative position among the sciences during the so-called 'golden age' of exploration from the fifteenth to the nineteenth centuries. This was not due to the academic status of the subject during this period, but to the work of a number of people, whom we may call geographers, who were actively involved with the mapping and description of the new lands which were being discovered. At that time the subject hardly existed as an academic discipline; there was little developed theory or method. A large number of people, however, received a basic training in cartography, especially during military service.

In nineteenth-century Europe, a steadily growing interest in the investigation of the remaining blank spaces on the map of the world and in the development of international trade and colonial power, led to the establishment of a number of geographical societies. Their founders were enthusiastic scientists and others who sought to widen the support given to research and to expeditions. The societies were supported by prominent members of the middle classes. They also received help from governments during the period of colonial expansion at the end of the century. Although forerunners of these societies existed during the sixteenth and seventeenth centuries, the first modern geographical society was founded as the Société de Géographie de Paris in 1821. In rapid succession came the Gesellschaft für Erdkunde zu Berlin in 1828, the Royal Geographical Society in London in 1830, then societies in Mexico (1833), Frankfurt (1836), Brazil (1838), The Imperial Russian Geographical Society in St Petersburg (1845), and the American Geographical Society in 1852. By 1885 nearly 100 geographical societies with an estimated membership of over 50 000 were spread across the world (Freeman 1961, pp52–3).

The most important work of these societies was their support for expeditions and their publication of yearbooks and journals which included maps and other material from expeditions. Many of the societies also supported colonial expansion by their respective countries. Some, such as the societies at Nancy and Montpellier, supported local studies in their home areas. At Nancy, barely 10 km from the Franco–German frontier of 1871, the Société Géographique de L'Est pursued studies supporting the return of Alsace–Lorraine to France.

During this period the geographical societies held an important public position because of their commitment to exploration and popular political causes. Their meetings attracted great public interest, especially when one of the well known explorers returned to give an account of his discoveries and adventures. Eventually all the blank spaces on the map of the world were investigated and explorers no longer brought back sensational stories. Today, discoveries in outer space make the greatest appeal to man's urge to discover the unknown. Space research, however, requires more highly specialised techniques than geographers or geographical societies can command; the collection and mapping of data about the surface of the Earth have been taken over by government organisations like the Ordnance Survey and the US Bureau of the Census. This activity has led to an explosion of information about the world so that the knowledge available to the modern geographer is many hundred times greater than his predecessor in the mid-nineteenth century could command. Further development of remote sensing and data banks will continue to multiply the amount of information during the next few years. It is perhaps not surprising that, confronted with these developments, the majority of geographers are now less involved in the direct collection of data in the field, and more concerned with the search for general principles and laws through systematic analysis of material collected by others.

Electronic computers and advanced statistical techniques have become essential for the handling of the growing flood of information about

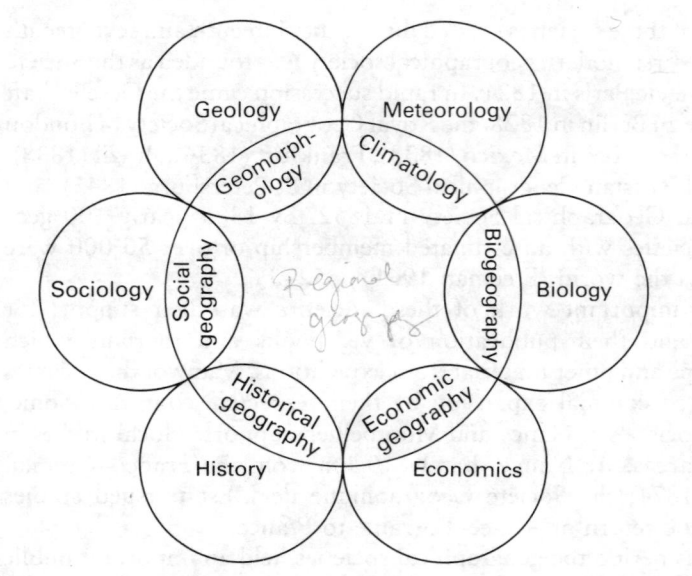

Figure 1. The circumference of geography. (Adapted from Fenneman 1919)

geographical phenomena. The use of these techniques is not, of course, confined to geography but is common to all sciences. For a science to play a distinctive and useful role, it needs both a specific field of enquiry and a set of specific concepts. What are the characteristics of geography in these respects?

Firstly, geography may be regarded as *a science of synthesis*. A student immediately notices that geography has no obvious place in the traditional classification of the sciences by faculty. Some parts of geography have their strongest affiliations with mathematics and natural sciences, others with history, philosophy and social sciences. Other sciences study distinctive types of phenomena: geologists study rocks, botanists plants, sociologists social groups, and so on. The work of geographers involves several types of phenomena, each already studied by another science. Are geographers, therefore, 'jacks of all trades and masters of none?'

Some geographers argue that the subject matter is shared with other disciplines but is treated in a different way for geographical purposes. Others affirm that the subject matter of geography is exclusive; geographers alone study places. All geographers will however agree with Ackerman (1958, p8) that the fundamental approach in geography 'is the differentiation of the content of space on the Earth's surface and the analysis of space relations within the same universe.'

To clarify this point we may analyse the position of geography as seen by Fenneman (1919) (figure 1). The diagram expresses the conception that sciences overlap and that each one of the overlapping zones, which also represent specialised, *systematic branches of geography*, belong equally to some other science. The question is whether geography occupies anything more than these overlapping zones. Is anything left in the centre which is specifically geographical? What would be missing if the geography of an area were written by a group of scientists who each contributed his own chapter: the geologist writing about the rocks, the botanist about the plant life, the meteorologist about climate, the demographer about population conditions and the economist about economic conditions? Such an account would surely fail to consider the overall interaction between phenomena. It is easy to see, for example, that the relationship between climate and soil type must have an important bearing on conditions of agricultural production and that the development of industry in an area may not be due only to economic factors but also to the natural resources of the area, its population potential and its historical and political development. Fenneman concluded that the interaction of all these factors can primarily be studied within definite areas or regions, and argued that geography should cultivate its core, *regional geography*, 'as a safeguard against absorption by other sciences.' Regional geography is defined as the study of areas in their total composition or complexity. In most cases, regional geography would, however, focus on the relationship between Man and his habitat or another theme that makes an illuminating presentation of the region possible.

While there is still some truth in the assessment that many geographers have lost their geographical identity to alien subjects when working on the periphery, Fenneman's fear of absorption by other disciplines seems rather strange today. Instead of illustrating the robbers trying to disintegrate our discipline, figure 1 may be used to demonstrate the point made by Ackerman (1958, p3), that geography is 'a mother discipline' from which other specialised disciplines like geodesy, meteorology, soil science, plant ecology and regional science have emerged. Geography has become an outward-looking discipline which has frequently created new specialisations. This multidisciplinary perspective may be regarded both as our *raison d'être* and our life-raft in the sea of knowledge (Capelle 1979, p65). If the periphery seems interesting, why not explore it; this will only widen the 'circle of geography.'

Minshull (1970, p28) considers that 'many of the subjects from which geography is said to borrow just do not exist.' The systematic branches of human geography especially, are breaking much new ground. Admittedly, useful knowledge from other disciplines can be fitted to the procedures used in geography but no other specific procedures are designed or followed in order to reveal the intricacies of space relations. The only other science concerned mainly with space relations is astronomy, which has a different universe to treat. Biology and economics also have treated space relations but within much more limited universes than geography (Ackerman 1958, p27).

Economic geographers, for instance, start with the spatial distribution of different forms of economic activity and try to explain this distribution, while economists are in general less interested in spatial distribution, preferring to concentrate on the factors which determine economic development on the macro scale.

There is even in general a difference between the botanist specialising in plant distribution and a geographer interested in vegetation. In most cases, the geographer will concentrate on vegetation types and their distribution, and will instinctively delimit the topic with its importance to human geography somewhere at the back of his mind. The botanist, on the other hand, will be more interested in the distribution of single plant species or grouping of species; vegetation as a part of the landscape picture and its importance to man recedes into the background, rare and inconspicuous flowers are often of greater interest to him than common and landscape-forming trees.

Geography exists in order to study variations in phenomena from place to place, and its value as an academic discipline depends on the extent to which it can clarify the connections between different features of the same area. The central geographical question is 'Why is it like this here?'

Peter Haggett (1979) illustrates the geographical viewpoint by discussing the starting points of different scientists who might all be studying the same beach full of people bathing and sunning themselves. The geologist would be interested in the sand particles and the zoologist in the marine life along the shore. The sociologist might study the behaviour of the groups using the beach and the economist might well be concerned with the marginal costs of the different ice cream salesmen.

The first reaction of a geographer would presumably be to map accurately the exact location of each activity. He would find it difficult to work on the ground and might want a general oversight, perhaps using a helicopter to cover the situation by taking air photographs. With a photograph the study area could be reduced to a scale which would be relatively easy to work with later. Most scientists like to enlarge their phenomena, geographers do the opposite by reducing the scale of complex phenomena on maps and photographs. As a second stage the geographer would try to systematise what he observed on the photograph or map in some sort of spatial order. He could, for instance, make a natural division of the beach, separating the tidal and wave-washed zone from the dry shore. Or he could divide the beach into zones of different population density. The third task would be to explain how spatial variations in density distributions came about. In that explanation a range of distinctive factors would be discussed, including natural factors like exposure to sunshine and shelter provided by the topography, as well as cultural factors such as the distance from restaurants, car parks and toilets.

All the phenomena which have a recognisable spatial distribution or can be shown on a map are basically of geographical interest, but many are of little significance. It is, however, impossible to say unambiguously which of the phenomena are of no significance. This has to be decided by the circumstances of each case. In most cases, for example, the distribution of Mormons in a region would have little geographical interest, but where, as in parts of the American West, their distribution becomes dominant or has contributed to the cultural and economic development of the area, then it becomes significant.

In research it is not always necessary to take account of all the factors of geographic interest. The research worker might restrict himself to the distribution pattern of only one geographical phenomenon and to try to find an explanation for it. He might try to make valid generalisations about the distribution of settlements in the landscape. He might postulate a close relationship between population density and the distance between towns. In this case he is specialising in settlement geography, a systematic branch not mentioned in figure 1.

The number of branches of systematic geography could be argued indefinitely. In 1919 Fenneman even included 'mathematical geography' which has for a long time been included in astronomy. Minshull (1970, pp16–29) limits the field of geographic enquiry to five topics of the physical environment and three major topics of human life. The five topics in physical geography are: rock type and arrangement, relief and drainage, climate, soils and natural vegetation. Oceanography exists as a sixth specialist study, but the topic 'ocean currents and productivity' is often excluded from geography textbooks. While there is more or less agreement on the content of physical geography there is much less consensus on the topics of human geography. In a general way we might, however, agree on the topics mentioned by Minshull (economic activity, social organisation and settlement), but some would prefer to substitute the word settlement with the much wider concept of *habitat*, meaning the total physical milieu in which men live, including buildings and other human artifacts, as well as vegetation and other environmental features. On this basis Hannerberg (1961; 1968, p12) divides human geography into three branches: economic geography, social geography (including population geography) and cultural landscape geography. Today this division might seem too conventional, but we will leave this question now for a final discussion in the concluding paragraphs of this book.

Many geographers have regarded regional geography as the core of the subject, viewing systematic geography as the area in which laws are formulated and regional geography as the field in which they are tested empirically. The culmination of regional geography then becomes the verification of geographical laws and the presentation of a *synthesis* of the physical and human phenomena within an area or region.

It has, however, been difficult for regional geography to fulfil this role in the field of research. In fact, research has largely been confined to the systematic branches of geography since the 1930s — at least in the English-speaking world; regional geography has maintained a rather stronger position in France and Germany.

The last two decades have been marked by a general debate on the content and method of geography. The traditional subdivisions have been strongly criticised. The central task of geography remains to act as a synthesising science. The new geographical synthesis, however, may have different theoretical bases than traditional regional geography. Before we enter this contemporary discussion, it is, however, necessary to look at the historical development of geography as a science.

CHAPTER TWO

THE FOUNDATION OF SCIENTIFIC GEOGRAPHY

Geography in the Ancient World

Interest in geographical problems and writings on subjects which we can recognise as geographical, began long before the introduction of the subject into universities. In fact it is difficult to imagine how there were ever people who did not think geographically, who never considered the conditions under which they lived, and never wondered how people lived in other places. In this sense geographical thinking is older than the term 'geography', which was first used by scholars at the Museum in Alexandria about 300 BC.

The ancient Greeks made the first major contribution to the development of the subject. Scholarly writers produced *topographical descriptions* of places in the known world, discussing both natural conditions and the culture and way of life of the people who lived there. The best known of these was Herodotus (*c*485–425 BC). He was first and foremost an historian, and is indeed often regarded as the father of history. There are also good grounds for calling him the founder of geography because he placed historical events in a geographical setting; some of his writings are truly geographical in character. He not only described geographical phenomena as, for example, the annual flow of the Nile, but also attempted to explain them. In addition, Herodotus incorporated older sources of geographical information, including the existing maps, and he identified and criticised his sources. In this way Herodotus handed on knowledge of earlier theories and descriptions which would otherwise have been forgotten.

Herodotus had no interest in those mathematical and astronomical problems which later became associated with geography — the measurement of the circumference of the Earth and the establishment of exact locations for places. He accepted the Homeric view of the Earth as a flat disc over which the Sun travelled in an arc from east to west. The belief that the Earth was a sphere was discussed in Herodotus' time but it was Aristotle, about a century after Herodotus, who produced observational evidence for a spherical form for the Earth. The scholars at the Museum in Alexandria were then able to establish the foundations for the calculation of latitude, longitude and the size of the Earth.

Eratosthenes (276–194 BC), who was chief librarian at the Museum, succeeded in calculating the circumference of the Earth with remarkable precision. He actually measured the distance from Syene (Aswan) to Alexandria and estimated it to be one-fiftieth of the circumference of the Earth by observations of the elevation of the Sun. Of equal importance was his development of systems of coordinates for the world, i.e. latitude and longitude, which he used to locate places and to measure distances. This made it possible for him to draw the first passably accurate map. Geographical order now replaced the casual and unsystematic location measurements of earlier times.

Eratosthenes' cartographical work was later developed by his students and successors at the Museum in Alexandria. Ptolemy (AD 90–168) wrote a major work in eight volumes which is now known as *Ptolemy's Geography*. The first volume explained the principles for calculating the dimensions of the Earth, its division into degrees, calculations of latitude and longitude, and a discussion of map projections. The eighth volume contained maps of different parts of the world. Other volumes included tables of latitude and longitude for 4000 places.

Although it was widely known how to calculate geographical latitude from the altitude of the Sun in Ptolemy's time, calculations had only been made for a handful of places. The calculation of longitude was then only possible by estimating the length of journeys from one place to another and so many of Ptolemy's locations for places were erroneous. His biggest mistake was to underestimate the size of the Earth, rejecting the almost correct estimate of Eratosthenes in favour of a reckoning made by Posidonius in about 100 BC. Posidonius reckoned the circumference of the Earth to be 180 000 stadii (a stadium can be taken to represent 157.5 m), against Eratosthenes' estimates of 252 000 stadii. Because calculations of longitude had to be based on rather unreliable travel distances, Ptolemy's otherwise remarkably accurate map of the known world included too many degrees of longitude. The map extended from a prime meridian in the Canary Islands to a 180 degree meridian which crossed inner China. The actual distance between these places is only 120 degrees.

The topographical tradition of Greek geography, represented by Herodotus, was carried forward into Roman times. Strabo (64 BC–AD 20) wrote a work of 17 volumes called *Geographica*. This was largely an encyclopaedic description of the known world whose chief value was that it preserved for posterity many writings which he annotated and cited. *Geographica* also included attempts to explain cultural distinctiveness, types of governments and customs in particular places. The significance of natural conditions for cultural development was discussed in relation to a number of places, especially in the description of Italy.

Middle Ages and Renaissance

The Middle Ages were a dark period for the development of science in Europe. At best, scholars made accurate but sterile copies of the works of the ancients, rejecting anything which did not conform with the dogmas of the Church. Such an intellectual environment stifled any development of critical scientific analysis. Concepts of the world which had been developed in ancient times were reshaped to conform to the teaching of the Church. The Earth became a flat disc with Jerusalem at its centre.

Ancient learning was, however, carefully preserved in the Islamic countries at the Arab universities of Spain, North Africa and the Middle East. Arab traders travelled widely and gathered information which could be used by scholars to fill in the gaps on Ptolemy's original map. The best known of these travellers was Ibn Batuta (1304–1368) who travelled as far east as Peking in China, and sailed far south of the equator along the east coast of Africa. This particular trip showed that Aristotle had been wrong in believing that it was too hot for human habitation in what the Greeks had called the 'torrid zone.' By the twelfth century, Al Idrisi had shown that the Greek division of the world into five climatic zones (two cold, two temperate and one torrid) did not correspond to reality and had suggested a more sophisticated world climatic system.

The last great Islamic geographer of the Middle Ages was Ibn Khaldun (1332–1406). He established the foundation for historical geography in those of his writings which analysed the rise and fall of empires. He suggested that warlike nomads often founded large states, but that after a while the nomads were absorbed by their permanently settled subjects. As peasants and townsmen, the rulers lost their warlike spirit and eventually their kingdoms fell apart. Ibn Khaldun both predicted and lived to see the collapse of the Islamic state he lived in. At the fall of Damascus in 1400 he actually met Tamerlane, the conqueror and devastator. It was unfortunate that the works of Khaldun and Al Idrisi were not translated into Latin or any other West European language until the nineteenth century; as a result, the Europeans were unable to make use of the observations of these Arab scholars (Broek 1965, p12).

The journeys undertaken by Europeans during the Middle Ages made little significant contribution to the development of geographical knowledge. Around AD 1000 the Norsemen sailed across the North Atlantic to Greenland and North America but the Sagas of these adventures were only passed on by word of mouth and written down long afterwards in isolated Iceland.

Meanwhile, exploration and learning flourished in China. Actually Europe and India were 'discovered' by Chinese travellers long before Europeans reached the Orient. In the period between the second century BC and the fifteenth century AD, Chinese culture was the most efficient in the world in applying knowledge of nature to useful purposes. The study of geography was advanced well beyond anything known in Europe at this time; among other things, the Chinese used coordinates and triangulation to produce beautiful maps of China and neighbouring countries. When, however, the Italian Marco Polo (1255–1325) wrote an account of his travels to China describing the high level of Chinese learning, his book was widely discounted as a fictitious adventure.

The Renaissance brought about a renewed interest in the geographical knowledge of ancient times. Copies of Ptolemy's *Geography* which had been preserved in Byzantium (Istanbul) were discovered and brought to Venice where they were translated into Latin in about 1410 and made a great impact on contemporary scholars (Bagrow 1945). Columbus and other expeditionary leaders relied on Ptolemy's calculations (p10), but some of their more famous discoveries disproved the latter's calculations of longitude and changed the world picture he had established. There were new developments in cartography — new projections, especially that of Mercator in 1569, were invented; the first globes were made, and new world maps were published.

In addition to this there was a revival in another branch of ancient geography, that of topographic description. Large numbers of accounts of voyages provided raw material for encyclopaedic works on the geography of the world. The jumble of information related to place names included in such *topographies* where notes on natural and physical conditions were juxtaposed beside miscellaneous information on folklore, etymology and history, seems to us to place them outside the field of science. The popularity of these works in their own day might well puzzle us today until we realise that we "are often prisoners of contemporary logic and cannot see those qualities of old works which cannot be integrated into our system . . . at the time it had just as much relevance as what continues to interest us today: it was part of what the episteme of the time indicated was knowledge." (Claval 1980 p379).

Varenius and Kant

Strabo made a rough definition of the study of geography when he said that, 'wide learning, which alone makes it possible to undertake a work on geography, is possessed solely by the man who has investigated things both human and divine . . . The utility of geography . . . regards knowledge both of the heavens and of things on land and sea, animals, plants, fruits, and everything else to be seen in various regions — the utility of geography, I say, presupposes in the geographer the same philosopher, the man who busies himself with the investigation of the art of life, that is, of happiness' (cited by Fischer *et al* 1969, p16). Strabo considered that geographers should study both human activities and natural conditions.

This dualism within the subject is still characteristic of geography. Some writers have regarded it as the essential justification for the role of geography, while others have argued for a division of the subject into physical and human geography on the ground that the respective methodologies of physical and human geography must be different. In studies of natural phenomena, including climate, geology and land forms, it is possible to use the methods of the natural sciences and to draw conclusions with a large measure of scientific precision. The methods of natural science, however, do not lend themselves very well to the study of social and cultural phenomena. Our generalisations about human groups must be limited in time and space, and must relate to statements of probability rather than certainty.

Bernhard Varenius (1622–50), whose *Geographia Generalis* was published in Amsterdam in 1650, was one of the first scholars to suggest these essential differences in the character of physical and human geography. His book includes much material which today would be treated as mathematical geography or as astronomy. The *Geographia Generalis* is divided into three parts: (1) the absolute or terrestrial section, which describes the shape and size of the Earth and the physical geography of continents, seas and the atmosphere; (2) the relative or cosmic section, which treats the relations between the Earth and other heavenly bodies, especially the Sun and its influence on world climate; and (3) the comparative section, which discusses the location of different places in relation to each other and the principles of navigation.

Varenius intended that the *Geographia Generalis* should be followed by what he called, in his foreword to the book, 'special' geography in which descriptions of particular places should be based upon (1) 'celestial conditions,' including climates and climatic zones; (2) terrestrial conditions with descriptions of relief, vegetation and animal life; and also (3) human conditions including trade, settlement and forms of government in each country. In fact he showed little enthusiasm to embark on the study of human geography as it was not possible to treat it in an exact way. He explained that he had included this last group of conditions as a concession to traditional approaches to the subject. Unfortunately, he did not have the opportunity to write on special geography since he died at the early age of 28 (Lange 1961).

Varenius made two major contributions to the development of geography. Firstly, he brought together contemporary knowledge of astronomy and cartography and subjected the different theories of his day to sound critical analysis. Secondly, his partition of geography into 'general' and 'special' sections originated what we now call systematic and regional geography. *Geographia Generalis* dealt with the whole world as a unit, but was restricted to physical conditions which could be understood through natural laws. 'Special' geography was primarily intended as a description of individual countries and world regions. It was difficult to establish 'laws' in the 'special' (regional) part for the explanation must be descriptive where people are involved. Preston James (1972, p124) points out that the general (systematic) and special (regional) parts of Varenius' work complement each other, and that Varenius saw 'general' and 'special' geography as two mutually interdependent parts of a whole.

The lectures on physical geography given by Immanuel Kant (1724–1804) in Königsberg were of great significance for the later development of geography. The course, which was given more than 40 times, was one of his most popular. It was not a course on physical geography as we now understand it, but also studied the races of man, their physical activities on the Earth and natural conditions in the widest sense of the term. For Kant himself geography represented only an approach to the empirical knowledge which was necessary for his philosophical research, 'But, finding the subject inadequately developed and organised, he devoted a great deal of attention to the assembly and organisation of materials from a wide variety of sources, and also to the consideration of a number of specific problems — for example, the deflection of wind direction resulting from the Earth's rotation' (Hartshorne 1939, p38).

Kant's great contribution to geography is arguably that he provided a philosophical foundation for the belief that the subject has a significant scientific contribution to make. He pointed out that there are two different ways of grouping or classifying empirical phenomena for the purpose of studying them: either in accordance with their nature, or in accordance with their position in time and space. The former is a *logical* classification, the latter a *physical* one. Logical classification lays the foundations for the *systematic sciences* — the study of animals is zoology, that of the rocks is geology, that of social groups is sociology, and so on. Physical classification gives the scientific basis for history and geography. History studies the phenomena which follow one after the other in time (*chronological science*), while geography studies phenomena which lie side by side in space (*chorological science*). History and geography are both essential sciences, standing alongside the systematic sciences. Without them, man cannot achieve a full understanding of the world.

Kant gave geography a central place amongst the sciences and geographers have often reiterated his views in justification of the existence of the subject and its special position amongst the sciences. Alfred Hettner (1859–1941) who did most to redevelop Kant's thought, admitted (1927, p115) that for a long time he had not paid sufficient attention to Kant's exposition of geography but had later rejoiced to discover a close correspondence between the conclusions of the great philosopher and his own. The American geographer Richard Hartshorne closely followed Hettner in his *Nature of Geography* (1939), considering that Kant's approach led towards a satisfactory understanding of the nature of geography and answered all its basic questions.

Today it is widely considered to be impossible, and to some extent philosophically untenable, to draw such sharp divisions between the 'sciences' as Kant did. The systematic sciences study phenomena with reference to time and space and it is very difficult to separate time and space in studies of human geography. Understanding of geographical situations is always improved when we consider their development over time. Just as there are geographers who wish to study the cultural landscapes of former times without necessarily using their knowledge to illuminate contemporary conditions, so historians cannot neglect the study of the distribution patterns of phenomena. Here the individual sciences overlap each other.

Hettner (1927, pp131–2) considered that the biggest difference between history and geography 'does not lie in that geography sets out to study a given time, namely the present, but that in geography the time aspect recedes into the background. Geography does not study development in time as such — although this particular methodological rule is often broken — geographers cut through reality at a distinct point in time and only consider historical developments in so far as they are necessary to explain the situation at that chosen point in time.'

During the period between the death of Kant in 1804 and Hettner's book in 1927, this 'methodological rule' had not only been broken but almost systematically opposed, since Kant's philosophical guidelines for the subject had been virtually forgotten until Hettner and Hartshorne resurrected them. In the early nineteenth century, and more markedly after 1859, under the influence of the new scientific thinking which came in with Darwinism, geographical research was closely associated with the study of process, i.e. the development of phenomena over time in geographical space. For the French school of regional geography, which is regarded as one of the most influential movements within the subject, development over time is a central concept for both method and presentation. The directions which were developed during this 'traditional period' provided a methodology for the subject, whereas the philosophical ballast presented by Hettner and Hartshorne for their successors provides an apparently solid philosophy for the subject but gives no clear methodological directions for geographical research. This point will be discussed in more detail in Chapter 3 (p52–5).

The Classical Period

Kant gave geography a philosophical foundation but Alexander von Humboldt (1769–1859) and Carl Ritter (1779–1859) developed the subject as an independent branch of knowledge. These two men had many views in common and were united in their criticism of the casual and unsystematic treatment of geographical data by their predecessors. The similarity of their opinions was not accidental— Ritter regarded himself as a student of Humboldt, and Humboldt described Ritter as 'his old friend.' They differed markedly, however, in temperament and character.

Humboldt was the last of the great polymaths. He mastered a number of disciplines and put all his energy (and fortune) into travel and research in order to understand the whole complex system of the universe. In his great work *Kosmos,* with its subtitle *Sketch of a Physical Description of the*

World, he attempted to assemble all the contemporary knowledge of the material world. *Kosmos* was published in five volumes at the end of his life (1845–1862), while the results from his long and important journeys in Latin America (1799–1804) had been published in 30 volumes in Paris (1805–1834). When describing countries and places he strongly emphasised comparative studies. This gave a scientific stamp to his work which had been sadly lacking in earlier geographical descriptions.

Humboldt combined this *comparative method* with exceptionally sharp observation, and in fieldwork he was unsurpassed. His books and letters demonstrate that nothing escaped his observation. This, together with his training in geology, is shown in a letter he wrote to a Russian minister after a visit to the Urals: 'The Ural Mountains are a true El Dorado,' he wrote, 'and I am confident, from the analogy they present to the geological conformation of Brazil, that diamonds will be discovered in the gold and platinum washings of the Urals.' A few days later, diamonds were actually discovered (Tatham 1952, p55). Humboldt was primarily interested in physical geography. He undertook a large number of altitudinal measurements during his travels and drew height profiles of the continents. He carried out basic studies of plant geography and also wrote regional geographies of Mexico and Cuba which are of interest to the present day (Schmieder 1964).

Ritter's recognition as a geographer dates from the publication of the first two volumes of his *Erdkunde,* a volume on Africa and a volume on Asia, in 1817–18. The second edition was published in nineteen enlarged volumes in 1822–59. In 1820 he was established as the first professor of geography in Berlin, on the initiative of leading politicians. Although Ritter travelled widely in Europe, he spent relatively more time in his study than did Humboldt. The two men complemented each other in that Ritter was chiefly concerned with studies in human geography, and stood somewhat apart from the rapidly advancing research front of the natural sciences. It would be wrong to describe him as a determinist (see p24ff) as some commentators have done in the past. He certainly wrote about the influence of natural factors on man but was equally concerned with man's impact on nature. He believed, as did Vidal de la Blache much later, that 'the Earth and its inhabitants stand in the closest reciprocal relations, and one cannot be truly presented in all its relationships without the other. Hence history and geography must always remain inseparable. Land affects the inhabitants and the inhabitants the land' (cited by Tatham 1952, p44).

This implied that the individual region or continent had a unity, a

Ganzheit (which may be translated as a 'whole'), which it was the task of the geographer to study. This entity was something more than the sum of its parts — more than the totality of topographical, climatic, ethnic and other circumstances. Ritter's views were shaped by his deeply religious outlook and by the accepted natural philosophy of his time. His ideas on the 'wholeness' of things were in accordance with the writings of the German 'idealist' philosopher Georg W F Hegel (1770–1831), whose attitudes amounted to an attempt to comprehend the entire universe, to know the infinite and to see all things in God (Chisholm 1975, p33). The scientific stance of Ritter was *teleological* (Greek *teleos* = purpose). Teleology seeks to understand events in relation to their underlying purposes. Teleological explanations are therefore often regarded as the opposite of *mechanical* explanations, where the phenomena and observations are understood as outcomes of prime causes— such as the 'laws of nature' (see p20). Ritter studied the workings of nature in order to understand the purpose behind its order. His view of science sprang from his firm belief in God as the planner of the Universe. He regarded the Earth as an educational model for man, where nature had a God-given *purpose* which was to show the way for man's development. He did not regard the shape of continents as accidental but rather as determined by God, so that their form and location enabled them to play the role designed by God for the development of man.

Ritter's approach to knowledge was strongly criticised later; Hettner (1927, p87) said that Ritter's views were in accordance with the spirit of his time and this reinforced their influence on his generation. This influence, however, was bound to decline when Darwin's *Origin of Species* inaugurated a new philosophy of science.

Ritter combined a basic teleological standpoint with a most critical scientific precision. 'My system builds on facts, not on philosophical arguments', he said in a letter. The collection of facts was not an end in itself; the systematisation and comparison of data, region by region, would lead to a recognition of unity in apparent diversity. The plans of God, which give purpose and meaning, could only be discovered by taking into account all facts and relationships in the world as objectively as possible.

Ritter's significance as a scientist lies in his thorough and critical study of sources and in his ability to systematise extensive material, and in these respects his work is in the same tradition as that of Humboldt. While earlier regional descriptions had consisted to a large extent of the accumulation of unsystematised data about particular places, Humboldt and

Ritter effected a clear structuring of such material and, through deliberate research into both the similarities and the differences between countries and regions, sought to compare the different parts of the world with each other.

Subsequent writers have commented on the differences between the religious and scientific outlooks of Humboldt and Ritter, and have somewhat exaggerated them. Both men developed the comparative method and also laid great stress on the unity of nature. They both believed that the ultimate aim of research in physical geography was to clarify this unity and, in this respect, were in accord with the idealistic philosophies of their time. Humboldt did not pursue *idealism* as far as Ritter, for his concept of the unity of nature was more aesthetic than religious. In this respect he had more in common with Goethe than with Ritter. Unlike Ritter, he saw no reason to explain unity and order in nature as a God-given system to further man's development. Humbolt was strongly engaged in the gradual development of natural science and his greatest contributions lay in the field of *systematic* physical geography. Many regard him as the founder of biogeography and climatology. Ritter was to a considerable extent a *regional* geographer.

The geography which Ritter and Humboldt represented was designated as 'classical' by Hartshorne (1939), because it dominated the foundation period of the subject and because its methods were uniform and simple. Both Humboldt and Ritter believed that science must be founded on the objective descriptions of observed facts rather than on logical propositions like those proposed by Kant; geographers should collect empirical data and should later shape order and association from this material in order to arrive at an *inductive* explanation.

Darwinism and the Growth of Natural Sciences

The classical period can be said to have ended with the deaths of Humboldt and Ritter in 1859. The following decade experienced a whole range of new directions in geography. The consensus which had characterised the subject before 1859 gradually disappeared, although the viewpoints of the school which Ritter had founded lived on.

Ritter had become more and more interested in historical development in his zeal to demonstrate that the Earth is a school for mankind, and Ritter's students concentrated on historical studies to an even greater extent. In many schools of geography, especially in France, the subject came to be very closely associated with history.

The most adventurous of Ritter's disciples was the French geographer and revolutionary anarchist Elisée Réclus (1830–1905) who leaned less towards history and teleological thinking than most of the others. He achieved recognition with a work of systematic physical geography called *La Terre* (1866–7), but is best remembered for his 19-volume regional world geography *Nouvelle Géographie Universelle* (1875–1894). The clarity and accuracy of this work made it much more popular than Ritter's *Erdkunde* which had been its examplar in many respects, but which also contained many obscure passages. Réclus' work became a model for a range of encyclopaedic studies of the geography of the world and of particular countries.

The year 1859, which saw the death of Humboldt and Ritter, was also the year of publication of Charles Darwin's *Origin of Species*. The foundations of Ritter's approach were gradually replaced by a materialistic scientific philosophy which emphasised natural laws and causality, mechanical rather than teleological explanation. Many scholars came to believe that Ritter's work was valueless and unscientific because it could not be accommodated in Darwin's concepts of struggle and survival, and to his belief that evolutionary change resulted from random variations. The new scientific method which came to dominate research was, in a way, the opposite of Ritter's inductive approach which sought to witness to God's plan and existence (*final causes*, the aim or purpose of things observed). The method of natural science or, more precisely, the hypothetic–deductive method, seeks *prime causes* by *deduction*. It sets up hypotheses to see whether they can explain the reality which is observed. The testing of hypotheses leads to the formulation of natural laws. It was agreed amongst scientists that religion could not provide explanations of natural phenomena; even some theologians asserted that the Bible was not authoritative on scientific questions. Discussion as to whether or not there was a thought or a God behind the laws of nature or whether natural laws were a means of development to a designed end, came to be regarded as unscientific. Science should concern itself with causes rather than purposes.

The increasing belief in the inherent value of science led to a considerable growth in research into the natural sciences. It was not Ritter's disciples, with their historical interests, who won a place for geography in academic establishments during the latter half of the nineteenth century, but a group of natural scientists with an interest in physical geography. In their research they were able to build on the earlier work of Humboldt and Mary Somerville, whose *Physical Geography* had been published in 1848.

In contrast with Humboldt and Somerville, later nineteenth-century geographers concerned themselves more and more with pure physical geography where man played no part. There was also a tendency to consider physical geography to be a branch of geology. This was reasonable to the extent that natural scientists with a background in geology were then making the running in the subject. They established *geomorphology*, the study of landforms, which later became the most substantial element in physical geography. The biological elements in nature characteristically played a relatively small role in the teaching of geography in Britain and North America, so that physical geography is a more appropriate description of what was taught than the old term 'natural geography.'

In Germany the new direction was led by Ferdinand von Richthofen and Oscar Peschel. Peschel published *New Problems of Comparative Geography as a Search for a Morphology of the Earth's Surface* in 1870. Like Ritter, he was interested in the 'significance of land forms for the historical development of mankind', but he did not share Ritter's religious outlook, being more concerned with causes and effects as illustrated by the methods of natural science. He believed that geography should be a systematic, empirical brand of knowledge with its main emphasis on the study of landforms.

Germany became a leading nation for the development of academic geography in the mid-nineteenth century and the subject became a permanent independent university discipline. Although there had been a number of chairs in geography before 1874, a decision by the Prussian government in that year to establish permanent chairs in geography at all Prussian universities was an event of major importance. It is reasonable to assume that Prussia took this step in the belief that geographical knowledge could be used to further the political expansion of the state. At any rate, this decision meant that geography was firmly established as an academic discipline in one of the leading European nations; by 1880 there were professors in geography in ten of the Prussian universities (James 1972, p217).

In Britain, the first personal chair of geography was held by Captain James Machonochie at University College, London, from 1833 until 1836 but permanent university teaching dates from the appointment of Halford J Mackinder to a readership at Oxford in 1887 and the establishment of the first British department there in 1900. In the United States, individual university teachers of geography included William Morris Davis who began teaching at Harvard in 1878; the first department was established at the University of Chicago in 1903. It is typical that Davis, like the majority of his professorial colleagues in other countries, was a physical geographer.

By this time physical geography had become the chief field of study for geographers. In Britain no work was more influential in changing the teaching of geography at all levels than Thomas Henry Huxley's *Physiography*, first published in 1877. At this time the term *geomorphology* had not yet been invented; only *morphology* was in general use. *Physiography* has a much wider meaning; it may be defined as a 'description of nature', encompassing the systematic sciences of zoology, botany and geology. Huxley's work gives a clear demonstration of the post-Darwinian causal reasoning. In the introduction to his book Huxley argued that when we start to question the reason behind a simple phenomenon in nature we find a first 'cause, which will again suggest another, until, step by step, the conviction dawns upon the learner that, to attain even an elementary conception of what goes on in his parish, he must know something about the universe; that the pebble he kicks aside would not be what it is and where it is unless a particular chapter of the earth's history, finished untold ages ago, had been exactly what it was' (cited by Stoddart 1975, p21). One reason for the success of Huxley's book was its demonstration of better and more interesting ways of learning, linked directly with the pupil's own experience. The book begins with the Thames at London Bridge and, working from the local and familiar to the unfamiliar, it ends with the Earth as a planet. The characteristic of the book is its emphasis on experimentation and local studies. The idea that geography can only be learned through local studies, field courses and excursions is to a large degree derived from the educational principles of Huxley.

Physical geography (renamed physiography after 1877) became a very popular school subject during the last third of the nineteenth century, accounting for some ten per cent of the examination papers sat in English and Welsh schools during that time (Stoddart 1975, p26). Physiography was now regarded as an integral, if not the most important part of geography. A lectureship in physiography and commercial geography was established at Heriot-Watt College, Edinburgh in 1886, and a chair of physiography at the City of London College was created in 1894.

As more specialised Earth sciences became established, physiography as a comprehensive subject eventually vanished from the syllabuses. Within geography, its place was filled by the new science of *geomorphology* (introduced in Britain in 1895). The American William Morris Davis (1850–1934) became the leading personality in the development of geomorphology, although he preferred to use the term 'physiography'. His organising principle was not simple causality, but rather the idea of regular change of form through time, as systematised in his scheme of the cycle of erosion. In Germany, Albrecht Penck made important contributions to the new subject.

The success of physical geography, be it termed as physiography, morphology or geomorphology, led the German Georg Gerland to suggest, in 1887, that the study of cultural phenomena should be separated from geography. He supported this view on the logical grounds that geography is the science of the Earth and that a science should be developed on the basis of natural scientific methods. Scientific methods, as defined, could not be used to study cultural phenomena.

A corresponding view was expressed to the Royal Geographical Society by the historian Edward A Freeman who could not understand how geography could be recognised as an independent university discipline when 'on one side a great deal belongs to history but on the other the geologists lay claim to much of its material' (Keltie 1886).

Developments in physical geography provided the main innovations in geography during the latter part of the nineteenth century. Gerland's statements indicated how far the pendulum had swung in this direction, and his arguments may have led many to conclude that geography had moved out of balance. Alongside the special influences imposed on geography by the rapid development of the natural sciences, it is also clear that the main ideas and research methods of Darwin left their mark on subsequent geographical research. Stoddart (1966) suggests that the following four main themes from Darwin's work can be traced in later geographical research: (1) Change through time or *evolution*, a general concept of gradual or even transition from lower or simpler to higher or more complicated forms. Darwin used the terms evolution and development essentially in the same sense. (2) Association and organisation — man as part of a living ecological organism. (3) Struggle and natural selection. (4) The randomness, or chance character of variations in nature. The concept of change over time is, for instance, found in Davis's cyclic system for the development of a landscape through the stages of youth, maturity and old age, which Davis himself described as evolution. The French school of regional geography also took up the study of evolution within the cultural landscape.

The ideas of association and organisation — which Darwin had inherited from earlier philosophers and scientists — have been rather tenacious in geographical research. The influential German geographer, Friedrich Ratzel (1844–1904) discusses in his *Political Geography* (1897), 'the state as an organism attached to the land.' Although this *organism* analogy was derived from the natural sciences, its roots may also be traced in the earlier idealist philosophy and the idea of *Ganzheit* or 'whole' as used by Ritter (p18–19). The *region*, that particular field of study for geographers, has been regarded as a unique functional complex, which, despite a steady stream of material and energy, is in apparent equilibrium and constitutes a 'whole' which is more than the sum of its parts. This understanding of the region is found in the French school of regional geography and it is also apparent in more recent university textbooks (Broek and Webb 1973, pp14–15). The idea of the region as an organic unity pervaded British geography in the first half of this century. Andrew John Herbertson (1865–1915) used the term 'macro-organism' for the 'complex entity' of physical and organic elements of the Earth's surface (Stoddart 1966, p691).

The concept of struggle and selection, which had parallels in the contemporary political ideas of both economic liberals and Marxists, is reflected in subsequent geographical research, although no agreement was reached as to whether variation and development occurred by chance or were determined. This reflects a certain ambiguity in Darwin's thought. In later revisions of his work Darwin abandoned the issue of random variation and chance, partly because he failed to discover the laws which he had earlier believed to govern chance variations, and partly as a response to churchmen who sought to reconcile evolutionary process with fundamental direction.

Environmental Determinism and Possibilism

Geographical research after Darwin was primarily concerned with recognising the laws of nature. In human geography this approach led to a rather deterministic view of struggle and survival. Nature was studied with open eyes, seeking as objectively as possible to identify the forces or processes which governed the formation of valleys, uplands and coastlines. A more restricted view was taken of human activity, the relationships between nature and man were considered to be of primary interest. Man's achievements were to be explained as consequences of natural conditions.

In this respect we are also indebted to the British philosopher Herbert Spencer (1820–1903) who elaborated what is known as *social Darwinism*. He believed that human societies closely resembled animal organisms and that human societies must struggle in order to survive in particular environments much as plant and animal organisms do. Politically, Spencer was a liberal, believing that the 'fittest individuals' would survive best in a free enterprise system. In his general law of evolution Spencer claimed that all evolution is characterised by concentration, differentiation and determination.

Friedrich Ratzel, widely recognised as the founder of human geography, was strongly influenced by such ideas. The first volume of his chief work, published in 1882, was entitled *Anthropogeography, or Outline of the Influences of the Geographical Environment upon History*. In this volume, which sought to develop the new methods of natural science within human geography, Ratzel stressed the extent to which men live under nature's laws. He regarded cultural forms as having been adapted and determined by natural conditions. Although he may be criticised today for his determinism, we should recognise that Ratzel broke new ground in demonstrating that cultural as well as natural phenomena can be subjected to systematic study. Before his time human geography had largely confined itself to regional studies.

Ratzel modified his environmental determinism in his later work. A different emphasis dominated the second volume of *Anthropogeography* (1891) which discussed the concentration and distribution of population, settlement forms, migrations and the diffusion of cultural characteristics. He did not merely explain phenomena in human geography in terms of natural conditions, but stressed the significance of the historical development and cultural background of populations. At one point he even declared (as quoted by Broek, 1965, p18): 'I could perhaps understand New England without knowing the land, but never without knowing the Puritan immigrants.'

In scientific circles, however, the first volume of *Anthropogeography* had a much greater impact than the second volume. Ratzel's American disciple Ellen Churchill Semple (1863–1932) in particular, laid great significance on the deterministic opinions of Ratzel in her teaching at American universities. One of her own major works was entitled *Influences of Geographical Environment* (1911). Ellen C Semple and others of the same way of thinking, including the geomorphologist William Morris Davis, came to dominate geography in America until the 1930s and had an influence over American school geography for much longer. The two most influential geographers who carried on the study of environmental influences until the middle of this century were Ellsworth Huntington, who related the rise of civilisation in the mid-latitudes and the lack of development in the tropics to climatic conditions, and Griffith Taylor, 'whose determinist views so angered politicians interested in the settlement of outback Australia that he was virtually hounded out of his homeland' (Johnston 1979, p33). Taylor moved to the USA in 1928 and later to Canada; he insisted that he was not an old-fashioned determinist, but based his views on scientific knowledge of the environment. In 1951 Taylor could state with satisfaction, 'Thirty years ago I predicted the future settlement-pattern in Australia. At Canberra (in 1948) it was very gratifying to be assured by the various members of the scientific research groups there, that my deductions (based purely on the environment) were completely justified' (1951, p7). At the age of 70, Taylor returned to Sydney, was welcomed as a national hero and was even commemorated with his picture on a stamp. This example shows that a scientifically based *environmental determinism* might have some merit, but we might easily react against such extreme statements as that by Sir Halford Mackinder (1887, p143) that 'no rational (human) geography can exist which is not built upon and subsequent to physical geography.'

Up to the middle of the twentieth century, school textbooks in most countries included strong elements of crude environmental determinism in their selection of topics. A type of commercial geography was popular which laid considerable emphasis on raw material supplies and little on the distribution of entrepreneurial skills and institutions. The influence of physical geography on transport routes was greatly overstressed and oversimplified.

The view that 'there are no necessities, only possibilities' was strongly urged by the French historian Lucien Febvre (1922), who termed this approach *possibilism*, and contrasted it strongly with environmental determinism. Febvre invented the term, but the development of the possibilist way of thinking started earlier, and is especially associated with the geographers Paul Vidal de la Blache and Jean Brunhes in France. Later, Isaiah Bowman and Carl Sauer became active advocates in the USA. The possibilists did not deny that there were natural limits to the activities of man, but emphasised the significance of man's choices of activity rather than the natural limitations to it.

In Germany the reaction against environmental determinism began earlier than in other countries. We have seen how Ratzel modified his earlier approach. Alfred Hettner (1859–1941) was the most significant contributor to the further development of possibilism. He asserted (1927) that the geographical synthesis is distorted when nature is regarded as dominant and man as subsidiary. Hettner believed that the unity of geography can only be maintained through Kant's concept of a chorological science — that is, a subject which studies things which are mutually coordinated, not subordinated, in space.

The French School of Regional Geography

Paul Vidal de la Blache (1848–1918) is regarded as the founder of modern French geography. He had the clearest insight into the weakness of deterministic arguments, realising the futility of setting man's natural surroundings in opposition to his social milieu and of regarding one as dominating the other. He considered it even less useful to tackle these relationships along systematic lines in the hope of discovering general laws governing the relationships between men and nature.

According to Vidal de la Blache, it is unreasonable to draw boundaries between natural and cultural phenomena; they should be regarded as united and inseparable. In an area of human settlement, nature changes significantly because of the presence of man, and these changes are greatest where the level of material culture of the community is highest. The animal and plant life of France during the nineteenth century, for example, was quite different from what it would have been had the country not been inhabited by men for centuries. It becomes impossible to study the natural landscape as something separate from the cultural landscape. Each community adjusts to prevailing natural conditions in its own way, and the result of the adjustment may reflect centuries of development. Each single small community therefore has characteristics which will not be found in other places, even in places where the natural conditions are practically the same. In the course of time man and nature adapt to each other like a snail and its shell. In fact, the relationship between man and nature becomes so intimate that it is not possible to distinguish the influence of man on nature from that of nature on man. The two influences fuse.

The area over which such an intimate relationship between man and nature has developed through the centuries constitutes a *region*. The study of such regions, each one of which is unique, should be the task of the geographer. Vidal therefore argued for regional geography and against systematic geography as the core of the discipline.

Vidal's method, which was inductive and historical, was best suited to regions which were 'local' in the sense of being somewhat isolated from the world around them and dominated by an agricultural way of life. These circumstances favoured the development of local traditions in architecture, agricultural practices and the general way of life; the communities lived in such close association with nature that they might be self-sufficient in the majority of goods. Vidal advised geographers to carry out research in folk museums and collections and to investigate agricultural equipment which had been used in the past in order to study the individuality of development of the region.

Vidal used the following illustration in order to underline the long associations between the major factors governing the development of a community (1903, p386): while the surface of a shallow lake is being swept by a gust of wind, the water is disturbed and confused but after a few minutes the contours of the bottom of the lake can clearly be seen again. In the same way, war, pestilence and civil strife can interrupt the development of a region and bring chaos for a while, but when the crisis is over the fundamental developments reassert themselves. Changes can come in such a community, and Vidal pointed out many developments which had taken place in the French regions during the centuries preceding the French Revolution. These developments had taken place within a stable framework of interaction between men and nature.

Vidal's characterisation fitted those communities in Europe with which he was most familiar fairly well. These had been agricultural societies throughout the Middle Ages and well into recent times, having essentially local interests and interaction patterns resulting from a long interplay between man and nature. In certain circumstances, and for certain members of the small upper classes, their contact field was national rather than local, but in general Vidal's concepts were appropriate. Ironically, however, he was describing a rapidly disappearing phenomenon. His method is still well suited to the study of the historical geography of Europe up to the Industrial Revolution and also for the study of those parts of the world where society depends on subsistence economies. As Wrigley (1965, p9) pointed out, however, Vidal's method is not so well suited to the study of regions which experienced the Industrial Revolution of the nineteenth century.

Towards the end of his career, Vidal was well aware of this situation. This is most evident in *La France de l'Est* (1917), in many ways his most original work, which studied the development of the landscapes and agricultural societies in Alsace and Lorraine over a period of two thousand years. A considerable portion of the book, which is arranged chronologically, is devoted to changes which preceded the French Revolution of 1789. The revolution produced a great ripple in the picture, but afterwards the main lines of development reasserted themselves. After a particular point in the nineteenth century, however (Vidal identified the year 1846 himself), the finely balanced interplay between man and nature was profoundly disturbed. The surface of the lake was, as it were, whipped up by something more significant and persistent than the earlier gusts. Eventually the visible contours of the bottom of the lake were reshaped. The building of canals and railways, together with the reformation of the Alsatian woollen industry, initiated the decline of the traditional local, self-sufficient economy. Industry was developed on the basis of new cheap and rapid means of transport and could mass-produce goods for a wider market. These developments reduced the value of the regional method in a growing number of areas.

Vidal regretted these developments, which he could not avoid observing. He considered that much of the best in French life was vanishing with the self-sufficient economy but his stature as a scholar enabled him to suggest possible new approaches for further research. In the future, he suggested (1917, p163), we should study the economic interplay between a region and the city centre which dominates it, rather than the interplay of natural and cultural elements. Despite the breakdown of the self-sufficient regional economy, Vidal's work has been and still is a great inspiration to a vital tradition in geography, that of the regional monograph.

Landscapes and Regions

Vidal de la Blache is only one, if the most significant, of the founders of regional geography. There are other directions and methods in regional geography, although their discussion and classification has been rather neglected by English-speaking geographers. In Germany and France, where regional geography has been regarded as the core of the subject until quite recently, contributions have been much more extensive in this field. German learning has a long tradition of classification and systematisation, and we can turn to Fochler-Hauke's *Geographie* (1959, pp 251–61) for a review of the different approaches to regional geography.

In doing so we must first clarify the concepts 'region' and 'landscape' and their German counterparts. The two German words *Land* and *Landschaft* may both be translated as 'region,' but *Land* is a definite unit — a county or a country normally defined by its administrative borders. *Länderkunde* is the art of describing such definite units, as in regional monographs like Demangeon's *Picardie* (1905) — regional geography in its traditional sense. The word *Landschaft*, literally analogous to the English word 'landscape' also means 'a scientifically defined geographical region' in German. A *Landschaft* may be either a specific unit area or type of area. *Landschaftskunde,* which concerns both the study of such small unique areas and the delimitation and classification of different types of regions, straddles regional and systematic geography. Hannerberg (1968, pp 132–3) considers that *Landschaftskunde* may be taken over by the systematic branches of geography, while *Länderkunde* remains as regional geography in its own right. With this background in mind we now turn to the five approaches to *Landschaftskunde* as distinguished by Fochler-Hauke: landscape morphology, landscape ecology, landscape chronology, regionalisation and landscape classification (or systematisation). The first three of these branches demonstrate three different approaches to the study of regions.

In the Fochler-Hauke classification, French regional geography would appear as *landscape chronology,* with a major concern for the development of regions over time. The same category includes a type of regional geography which was developed in North America between the wars, described by Derwent Whittlesey (1929) as *sequent occupance.* Sequent occupance studies the ways in which each culture uses a region in its own way. This is demonstrated in America where most regions experienced a sudden change from the Indian to the European culture, and in many parts of Europe which progressed from agrarian to industrial cultures. Sequent occupance stresses the stages in the development of a region, not, as in the France of Vidal de la Blache, through studies of local differentiation as a result of the long continued and largely undisturbed interplay of man and nature over the centuries, but by emphasising how easily shifts in regional character can take place.

The concept of *landscape ecology* is less familiar, but it represents a functional approach in regional geography. This functional approach has grown up to study the relationships between centres and their *Umland*, that is, the well developed theme of functional or centred regions. Landscape ecology is also concerned with the relationships which occur within a region in terms of transport, contact fields, etc.

Clearly, a functional approach does not need to restrict itself to the study of centred regions, even if this has come to be a most important theme in recent geography. It may also study the total interplay between the factors in an ecosystem, whether dominated by man or nature, as will be further discussed in the section on ecosystem analysis (pp117–26).

Landscape morphology is a form of regional geography which was particularly well developed in Germany between the wars. Otto Schlüter, who played a central role in its development, asserted in as early as 1906 that geographers should consider the form and spatial structure created by visible phenomena on the surface of the Earth as their unifying theme.

All human distributions of non-material character, such as social, economic, racial, psychological and political conditions, are excluded from this study as ends in themselves. The French scholar, Jean Brunhes, illustrated this view when he said that we should study the Earth as if we were sitting in a balloon and looking down upon it. We should analyse the landscape and the characteristic interplay of observable phenomena there. The visible landscape is a result both of natural conditions and forces and a manifestation of the work of man. The landscape itself creates a synthesis, there is no longer a gap between physical geography and the geography of man, both have the same object— the visible landscape— and there is also close contact in terms of methods. Because of this, says Leo Waibel (1933), the landscape morphological approach represents an advance for geography, although the field of enquiry is thereby greatly narrowed.

Landscape morphologists argued as to how much their field of enquiry should be limited. Should they consider, for example, man's movements and the transport of goods and services? Many landscape morphologists considered that these activities should not be objects of study, they might only be considered as explanatory factors in so far as they contribute to an understanding of the evolution and character of the landscape. According to Schlüter, however, the cultural landscape includes not only the routes and route patterns, but also the men and the goods which move along them.

Alfred Hettner was opposed to Schlüter's limitation of geographical study to the visible landscape. He was (as described by Dickinson 1969, p132) 'concerned with the uniqueness of areas, whether this uniqueness was evident in the visible landscape or not. He recognized the focal interest of landscape, but refused to recognize the limits set by it on the study of the human facts in space.'

Despite this disagreement, landscape morphology can be seen as an attempt to study regions on the basis of Hettner's principles. Hettner had provided the German geographers with a philosophical basis for geography as a distinct discipline but no clear-cut practical methods of enquiry. Schlüter provided a practical approach which proved to be useful in field studies everywhere. The landscape morphologists agreed with Hettner on the main philosophical issue, that geography is a chorological science which should focus on regional synthesis and only delve into historical developments to the extent necessary to explain the contemporary situation.

Landscape morphology had its widest following on the European continent. In the USA Carl Sauer gave a basic evaluation of the new direction in his *Morphology of Landscape* (1925) but, apart from this, landscape morphology attracted little interest in the New World. In 1939 Robert E. Dickinson described landscape geography as the most important new line of growth within the subject and blamed its lack of development in Britain on lack of interest in it by British scholars. In fact, landscape morphology was only developed by a limited number of British geographers although it provided an objective for several courses in cartography and in the interpretation of aerial photographs.

There has been an increasing interest in the cultural landscape in recent years, perhaps as a reaction against the uniformity of most modern buildings and other landscape features. The ease with which a landscape may be disfigured through modern technology has awakened a concern for the care of landscape. The uncontrolled building of houses and cabins has created so many eyesores that an active policy of landscape formation has become a recognised need. This concept includes both aesthetic and economic aspects. J.H. Appleton (1975) and others have demonstrated the importance of an educated perception of landscape if many of the attractions of long-settled countries like lowland Britain are to be preserved. The rapid growth of industrial archaeology, especially as a hobby of educated laymen in Europe, also illustrates a persistent general belief in the importance of preserving something of the visual culture of the past. Robert Newcomb (1979) calls for a 'planning of the past,' and gives suggestions for active use of historic features, in order to make a preservation of at least relics of historic landscapes and townscapes possible.

The demand for area planning in districts, counties and regions has also increased interest in studies of the cultural landscape. In Norway, for instance, the book *Mountain Regions and Recreation (Fjellbygd og feriefjell,* ed Sömme 1965), the result of research into the effects of the recreational use of upland areas, is an example of a modern landscape study which attempts to provide a sound basis for new planning laws. This, and many other landscape studies, may not, however, be classified as regional geography since only limited aspects of the geographical phenomena in the area or region in question are scrutinised. Landscape morphology was intended by Schlüter as an approach to regional geography, but, as time went on, a number of systematic studies of restricted features in the landscape emerged, such as morphological studies of house types and their distributions, village types, and so on. We might agree with Hannerberg (see p8) that cultural landscape geography is a systematic branch of geography, just as landscape morphology represents an approach to regional geography. In many cases it might be difficult to decide exactly what is regional and what is systematic geography.

Regional Studies in Britain

'There is fully as much confusion over the meaning of the words "regional studies" as there is over the German word *Landschaft*,' states James (1972, p 267), with special reference to British geography. *Regional studies* have at least three different connotations in Britain. Firstly, there are regional studies which amount to descriptions of segments of the Earth's surface, broadly synonymous with regional geography or *Länderkunde* as we have defined it above. Secondly, there are regional studies which seek to divide the surface of the Earth into either homogeneous or functional areas of varying size, which we have termed *regionalisation*. Thirdly, regional studies may denote regional specialisation, when an individual geographer devotes a large part of his life to study different aspects of some part of the world (a good example is Percy M Roxby of Liverpool, who built up a formidable expertise on China).

Regional studies in Britain were influenced by Vidal de la Blache and other French regional geographers, but to an even greater extent by the French sociologist Frédéric Le Play (1806–82) through his influential Scottish follower Sir Patrick Geddes (1854–1932). As a starting point for the study of social phenomena in different parts of the world, Le Play carried out research on family lifestyles and family budgets. He recognised that family life depended on the means of obtaining subsistence, that is, work, while the character of the latter is largely determined by the nature of the environment, that is place. This leads to the famous Le Play formula which is basic to his ideas — place, work, family — which Geddes transformed into the slogan *place, work, folk* as basic concepts in the study of cities and regions.

Although not a geographer himself, Geddes had a major influence on British geography, especially in the fields of regional survey, regionalisation and applied geography. Field study — observation and recording in the field — was basic to Geddes' teaching. This led on to what he called 'regional survey,' embracing place, work and folk (alternatively described as geography, economics and anthropology, or as environment, function and organism (Dickinson 1969, p204). Geddes and his followers established the Le Play Society, which grew out of the Sociological Society in order to foster regional surveys. Geddes saw an important application of regional survey in _regional planning_, comparing the regional surveyor to an old family doctor who could interpret specialist knowledge and apply it to the actual condition of the individual patient, which he would know intimately. As a man of radical views, Geddes was rather impatient with geographers who defined their subject as a _descriptive science_, which tells us what is. Geography should rather be an _applied science_ which tells us what ought to be (Stevenson 1978, p57). In this way, visionary planners — and we must mention Ebenezer Howard with his models of suburban land use and functional structure which were developed in order to plan garden cities (Howard 1901) in this connection — were exercising a significant influence on geographical thought in Britain before World War I, laying the foundations for studies in applied geography which have continued ever since.

Geddes also influenced the study of regionalisation. Andrew J Herbertson (1865–1915), who was an assistant to Geddes in Dundee, and later taught geography at Oxford, presented a scheme for a division of the world into natural regions, based on an association of surface features, climate and vegetation. A more direct application of Geddes' ideas occurs in the work of Charles B Fawcett (1883–1952). In his book _The Provinces of England_ (1919), which attracted considerable attention, he proposed a federal structure of twelve autonomous provinces for England. 'All six principles of the division were clearly inspired by Geddesian regionalism, particularly so in the organisation of each province round a "definite capital which should be the real focus of regional life," that provincial boundaries should "be drawn near watersheds, rather than across valleys," and that "the grouping of areas must pay regard to local patriotism and tradition" ' (Stevenson 1978, p59). Fawcett translated the somewhat obscure ideas of Geddes into a workable form, and made one of the first identifications of _functional regions_. It is also interesting to note that the use of the catchment area of a stream (which served as an emblem for Le Play Society), as a basis for regional divisions is still regarded as rather versatile by modern geographers (see Haggett 1979, pp61–2). The main translator of Geddesian thought into geography, Herbert J Fleure (1877–1969) found, however, that the grand systematisation of society and environment into 'valley sections' was too rigid. Nevertheless, by expanding, qualifying and developing Geddes' ideas, Fleure rendered them accessible and acceptable to geography. Like Herbertson, he also worked on the definition of global scale regions.

One concept derived from the work of Geddes was that of the regional survey of potential land quality and land use as a basic input to plans for economic development. On his own initiative, L Dudley Stamp (1898–1966) organised and directed the first British Land Utilisation Survey during the 1930s, employing some 22 000 school children in the mapping of land use on a scale of 1/2500 in their home district under the supervision of school and university teachers. When World War II began in 1939 the vital importance of the maps was quickly appreciated as essential data for the programme of expansion of food production necessitated by the German blockade. At the International Geographical Congress in Lisbon (1949), Stamp suggested the establishment of a World Land Use Survey. The idea was adopted and an international commission under the IGU (the International Geographical Union) was set up to supervise the work.

The first Land Use Survey eventually formed the basis for an official agricultural regionalisation of Britain. The economic depression of the 1930s accentuated the effects of longer-term economic changes which were creating marked divergences of economic prosperity and distress between the subregions of Britain. Geographers contributed to national and local studies of these prosperous and 'depressed' areas — although description predominated over analysis at this stage — and, during the 1940s, were widely involved in the planning of post-war reconstruction. Several other regionalisation projects sought to delimit single-attribute regions such as industrial, climatic, vegetational, morphological and social regions. 'Each had its links with the relevant systematic sciences — social geography with sociology, for example — and the key differentiating factor between the two was the geographers' focus on the region, the single-attribute region of his specialism and the multi-attribute region in the synthesis of his work with that of others to produce regional geographies' (Johnston 1979, p36). Whereas much effort between the wars was put into the development of methods of defining multi-attribute regions, systematic studies in various specialisms gradually claimed the attention of the great majority of British geographers.

Even those who regard all forms of regional geography as blind alleys concede that they will continue to exist within geography. E A Wrigley (1965, p13) put it this way: 'The regional period of geographic methodology like the "classical" (including determinism) has left many traces, some of which will perhaps prove permanent, on the methods used in organising and presenting geographical material. Any discipline is both the product and the victim of its own past successes and these were two of the most important successes thrown up by geographical scholarship.'

The widespread hold of regional geography in the past has discouraged many geographers from seeking general relationships and theories, and has led them to decry the formulation of geographical laws and models. Rejecting the general theories of the determinists, they sought refuge in regional methodologies where each area is unique and somewhat exceptional and must be studied as such. Many still consider that geographers should continue with *idiographical* methods, i.e. the description of unique phenomena and unique regions.

Amongst this regional school it has been suggested that geography is as much an art as science. Geographers, in common with other 'arts' scholars, should emphasise literary quality. Jan Broek (1965, p21) puts it this way: 'The humanities stress real persons and cases rather than models, quality rather than quantity, evaluation and evocation rather than calculation, beauty and wisdom rather than information. Geography shares these attitudes to some extent.'

The idiographic method and the humanistic way of thinking within the subject began to be strongly criticised in the 1950s. A new and dynamic school was developed and new methods brought into use. There was much talk of paradigm crises and revolution. We shall see what this implies in the next chapter, which attempts to bring together the threads from this chapter and to present 'the quantitative revolution' in its historical connection.

CHAPTER THREE

PARADIGMS AND REVOLUTIONS

'Geography is concerned to provide accurate, orderly, and rational description and interpretation of the variable character of the earth surface' (Hartshorne 1959, p21).

'A traditionally held view — that geography is concerned with giving man an orderly description of his world— makes clear the challenge faced by contemporary geographers . . . The contemporary stress is on geography as the study of spatial organisation expressed as patterns and processes' (Taaffe 1970, pp5–6).

'Geography can be regarded as a science concerned with the rational development, and testing, of theories that explain and predict the spatial distribution and location of various characteristics on the surface of the earth' (Yeates 1968, p1).

While geographers would be in general agreement that major changes took place in every aspect of geographical thought during the 1950s and 1960s, there is no general consensus on the significance of the many developments which took place in those decades, nor on their effect on the future of geography as an academic discipline. The above quotations illustrate the widely divergent views expressed during this period.

It is clear that Hartshorne and Yeates differ on the methods rather than the content of geography. While both are concerned with the variation of phenomena over the surface of the Earth, Hartshorne visualises geography as an idiographic science with its main emphasis on the description and elucidation of individual phenomena because they are unique; Yeates considers geography to be a *nomothetic* (law-giving) science which requires the development and testing of theories and models through hypothetic–deductive methods in order to develop geographical laws (pp20 and 46).

Taffe implies that geography had changed by 1970, or was in the process of changing from an idiographic to a nomothetic science. Explanatory models which were once thought to be satisfactory were discredited by a large number of geographers. John D Adams (1968, p6) said that 'geography is currently in the throes of a paradigm crisis. Instead of asking the traditional question "Is it geography?" or "What is geography?" geographers are now asking "What should geography be?" If a satisfactory answer is not found to the latter question the next question is likely to be "Is geography relevant?" '

It is clear from Chapter 2 that this was not the first crisis phase in the development of geography. Ritter's teleological framework did not satisfy the determinists: views which were scientifically acceptable to Ratzel and Semple were too deterministic for Hettner and Hartshorne.

Kuhn's Paradigms

Perhaps Thomas S Kuhn (1962, 1970) is right to claim that science is not a well regulated activity where each generation automatically builds upon the results achieved by earlier workers, but a process of varying tension in which tranquil periods characterised by a steady accretion of knowledge are separated by crises which can lead to upheaval within subject disciplines and breaks in continuity. While it is possible to determine objectively whether an explanatory framework is satisfactory and reasonable *within* a specific scientific tradition, we must choose *between* different scientific traditions, and this choice is subjective. We must select what Kuhn calls *paradigms* (models or exemplars) for scientific activity.

Kuhn defines paradigms (1962; 1970, p viii) as 'universally recognised scientific achievements that for a time provide model problems and solutions to a community of practitioners.' Haggett (1979, p21) defines them as 'a kind of supermodel. It provides intuitive or inductive rules about the kinds of phenomenona scientists should investigate and the best methods of investigation.' A paradigm is a theory of scientific tasks and methods which regulates the research of most geographers, for example, or, where there is conflict between paradigms, by a group of geographers. The paradigm tells researchers what they should be looking for and which methods are, in this case, 'geographic'.

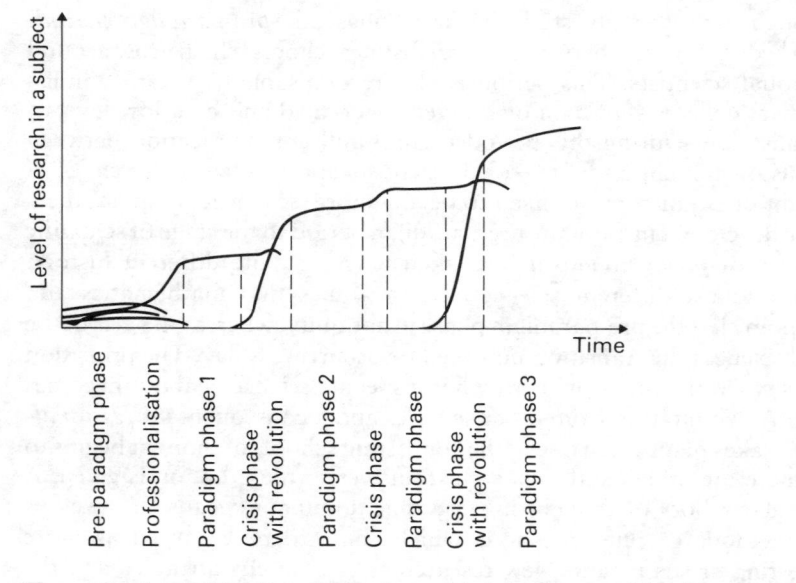

Figure 2. A graphical interpretation of Kuhn's theory of the development of science. (After Henriksen 1973)

A paradigm provides a scientific community with criteria for the choice of problems which, when the paradigm is taken for granted, can be assumed to have solutions. These usually become the only problems which the community concerned will recognise as scientific and encourage its members to take up. Other problems, including many which had previously been considered to be relevant, are rejected as either metaphysical, or as lying within the field of another science, or as being too complex to be worth spending time on. A paradigm can isolate a scientific community from significant problems which cannot be solved within their framework of its accepted principles and methods. Such problems bypass current scientific activity or what may be termed *normal science*, and therefore threaten the ruling paradigm. 'One of the reasons why normal science seems to progress so rapidly is that its practitioners concentrate on problems that only their lack of ingenuity should keep them from solving' (Kuhn 1962; 1970, p37).

The development of a science consists, according to Kuhn, of a series of phases (figure 2). A branch of science, beginning as a comparatively restricted philosophical problem region, becomes the subject for more thorough and systematic study. This first phase, the *pre-paradigm period*, is marked by conflicts between several distinct schools which grow around individual scientists. This period is also recognisable by a rather indiscriminate collection of data over a very wide field and by a low level of specialisation. During this period there is full communication between schools of thought and with other scientists and laymen. One school of thought does not consider itself to be any more 'scientific' than another.

The development from the pre-paradigm period to the stage of scientific maturity or *professionalism* has taken place at quite different historic dates amongst different sciences. Kuhn argues that mathematics and astronomy left the pre-paradigm phase in antiquity, whereas in parts of the social sciences the transition may well be occurring today. The transition begins when the question as to what a specific science is about becomes acute. A delimitation from other sciences and a consequent *professionalisation* takes place when one of the conflicting schools of thought begins to dominate the others and thus a clear answer to the question is given. A particular school of thought may become dominant because it develops new methods or puts questions which come to be regarded as more interesting or significant. New researchers are thereby attracted and the research makes progress. Earlier observations are re-interpreted and assimilated into a new unified framework.

A paradigm phase is characterised by a dominating school of thought which has, often in quite a short space of time, supplanted others. A paradigm is established which leads to concentrated research within a clearly distinguishable problem area, an activity described as 'normal science.'

To bring a normal research problem to a conclusion is to achieve anticipated results in a new way; Kuhn called this *puzzle-solving* because of its similarity to the solving of jigsaw or crossword puzzles. Although the paradigm does not provide any clearly spelled-out rules for problem solving, it provides an evaluation of the results rather than a blueprint for the method. The perception of the research worker is, in fact, constrained by his paradigm so that his observation of data is attracted to the expected result. There is an in-built opposition to unexpected discoveries. The paradigm can advance research and develop research economy in that the research worker can immediately reach the research frontier without having to define his philosophical bases and underlying concepts. The group identity of research workers also provides a strong psychological advantage which itself can stimulate scientific productivity. Only an exceptional man can undertake effective research on problems which are generally agreed to be valueless or uninteresting.

A period of 'normal science' is sooner or later replaced by a *crisis phase*. This occurs because more and more problems are accumulated which cannot be solved within the framework of the ruling paradigm. Either more observations shake the underlying theory or a new theory is developed which does not accord with the stipulations of the ruling paradigm.

The crisis phase is characterised by a re-assessment of former observational data, new theoretical thinking and free speculation. This involves basic philosophical debates and a thoroughgoing discussion of methodological questions. The crisis phase ends when it appears either that the old paradigm can solve the critical problems after all, allowing a period of normal science to be resumed, or that no significantly better theory to solve the problems can be developed and that, consequently, research must continue for a further period within the old paradigm. Otherwise the crisis phase ends when a new paradigm attracts a growing number of researchers.

The acceptance of a new paradigm inaugurates a *revolutionary phase*. This involves a complete break in the continuity of research, with a thoroughgoing reconstruction of the theoretical structure of a research field rather than a steady development and accumulation of knowledge. The understanding of truth itself and the scientist's perception of the world can take on a new dimension. The acceptance of a new paradigm is also revolutionary because it attracts the allegiance of the younger research workers in opposition to established scientists. The new scientific 'reason' seldom triumphs by convincing its opponents, but succeeds as they die and a new generation takes over. Younger workers who do not conform to the newly accepted paradigm are ignored by its followers. An evaluation committee, even when trying to be objective, is still constrained by the paradigm of its members and may block promotion for those who do not fall into line. During the crisis phase there is a period of opportunity for all groups but the battle is all the more difficult to live with. Researchers are continually forced to ask themselves whether the type of puzzle-solving they are doing is the 'right' one.

The exchange of one paradigm for another is not a wholly rational transaction. The new paradigm will generally provide solutions for the problems which the old one found difficult to resolve, but may not answer all the questions which were fairly easy to solve before. It is seldom possible to argue from logic that the new paradigm is better than the old. Even if a new paradigm can buttress itself with empirical and logical proofs, its original choice was basically subjective — an act of faith. Aesthetic considerations may influence the choice of a new paradigm, which may be regarded as simpler or more beautiful. It may give its adherents a revolutionary experience, shedding new light over a dark problem area.

Kuhn's picture of scientific activity is not wholly attractive. Our faith in the objectivity of research is weakened when we consider how subjective the choice of paradigms can be and experience the often protracted opposition of some scientific workers to the establishment of new explanatory models. Few research workers actually welcome the idea of a general debate on the value of research as an activity, in case it may lead to the evaporation of respect and financial support for research. On the other hand, as suggested by Peter Taylor (1976, p132), the youngest research workers at the bottom of the very formal academic hierarchy have a clear vested interest in changing the existing scientific ideology and thereby taking over from their elders.

Kuhn's model has given the new prophets a very effective weapon against the paradigm of a scientific 'establishment'. They do not need to justify their research as objective in itself; it is enough if they declare it to be objective within the subjective framework they have chosen. This can give rise to particular conflicts amongst social scientists because it is easy to equate the choice of a paradigm with the adoption of a value judgment. One could, for example, choose to work on the assumption that the 'Aryan' race is genetically and culturally superior to other races or that only a Marxist revolution will bring about a better world. These premises can form the basis for serious research but they will also influence the collection of data and the research approach in such a way that the results of the enquiry will support the ideologies. The ultimate conclusion may be that only those who affirm the same general outlook on the world and have similar political beliefs are competent to evaluate a piece of scientific research.

This brings us straight to the problem of social research and value judgments — postivism as opposed to critical philosophy — which will be discussed in Chapter 4. Here we can conclude that Kuhn's theories have had a positive influence on modern science in that they facilitate the acceptance of new theories and frameworks of understanding which may widen our knowledge and perception, but may have a negative influence in giving well organised groups of poorly qualified people a legitimate entry into research. There is a problem in universities where the number of students registering for specific subjects is seldom proportional to the theoretical development of the subject. Disciplines which offer simple and popular theories have a special attraction to incoming students.

How far does Kuhn's own theory provide such a simple solution, whose very clarity comes as a revelation to many but represents an incomplete or even false paradigm itself? Kuhn has developed a long-awaited new paradigm for the philosophy of science — for his book has been a best-seller — but it has the same advantages and disadvantages as other paradigms. It answers some but not all questions. His theory gives useful guidelines for the understanding of the historical development of a subject, but not a complete explanation. Kuhn was himself trained as a physicist and built his theory, in the main, on a study of the history of physics. How far is the history of physics relevant to less theoretical and quantitative sciences? To answer this question we must look at the history of geography in the light of Kuhn's theories.

Changing Paradigms in Geography

Figure 3 attempts to systematise the theoretical development of geography. It gives an incomplete and oversimplified picture, as only the main concepts in the development of the subject have been shown. These concepts have changed in significance and connotation over the course of time. In Kuhn's terminology (figure 2), geography was in the pre-paradigm phase until the time of Darwin. Kant did not found a school of geography but clarified the role of the subject and its position in relation to other sciences. Geography was, for him, a chorological and descriptive science distinct from the systematic sciences and from history. In his view, 'every foreign experience is transmitted to us either as a tale or a description. The former is history, the latter geography' (Rink 1802).

Ritter did not emphasise the distinctive roles of geography and history, but studied developments over time, linking history with geography in his work. In this respect Ritter and his school disagreed with Kant. Ritter was, however, the first geographer to give a clear description of his method. His account conforms to Francis Bacon's classical model of how a scientist works (figure 4A). A scientist starts with a range of sense-perceptions which he works up conceptually and verbally into a number of loosely arranged concepts and descriptions which we like to call facts. Next, certain definitions are necessary in order to organise the data. Afterwards the facts are evaluated and arranged in relation to the definitions.

The ordering and classification of data is often the chief activity of science in the early stages of its development. These first classifications

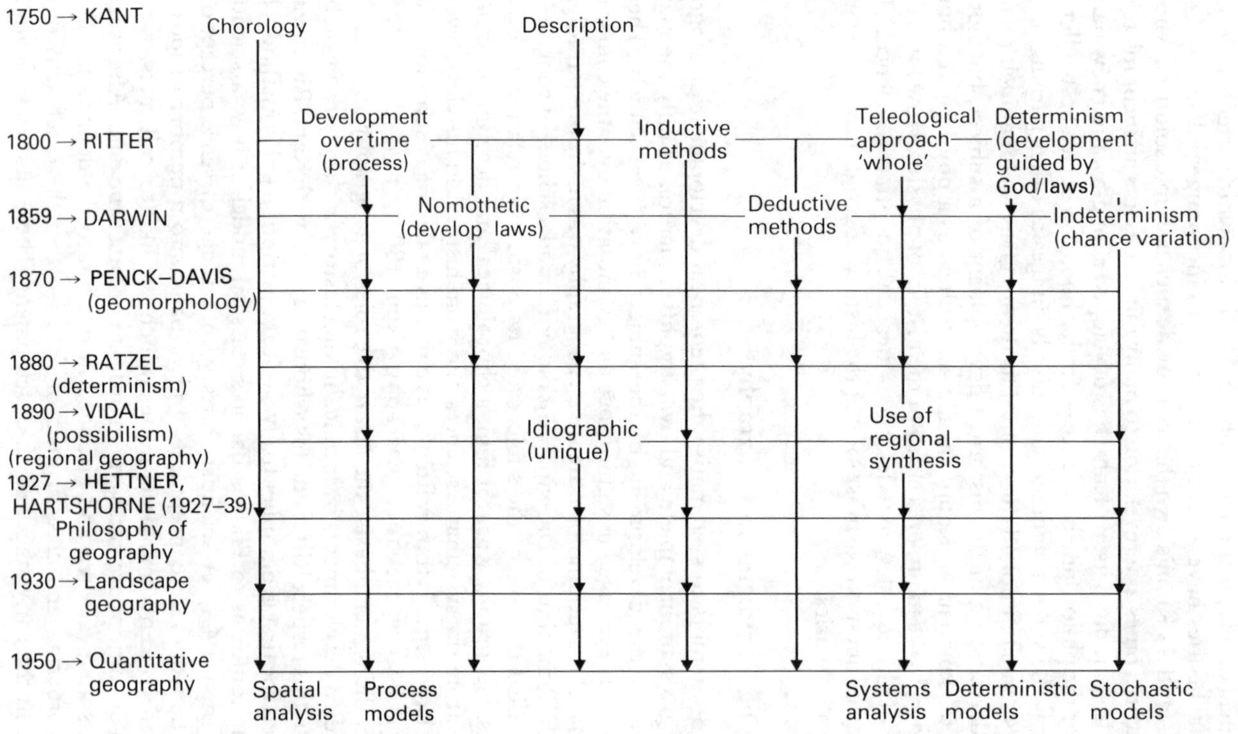

Figure 3. Ideas in geography 1750–1950—two centuries of development.

may have only a weak explanatory function. Continuing study of the interaction between classes and groups of phenomena reveals a number of regularities; such regularities and laws may be called *inductive laws* since they are derived from the observations of a large number of single instances.

Here we must clarify what a *scientific law* is. Braithwaite (1953, p12) defines a law as 'a generalisation of unrestricted range in time and space', in other words, a generalisation with universal validity. With this definition we can distinguish between empirical generalisations and laws. An empirical generalisation is valid for a specific time and place but a law is universal. James (1972, p473) maintains that a law within Braithwaite's rigorous definition can hardly be formulated on the basis of geographical evidence. The only truly universal laws are those of physics and chemistry, although even in physics there are elements of uncertainty which make probability calculations necessary. Harvey (1969, p31) gives the concept of law a much wider significance and postulates a threefold hierarchy of scientific statements from *factual statements* or systematised descriptions, through a middle tier of *empirical generalisations* or laws, to *general* or *theoretical laws*. This chapter uses the expression 'law' in the wider sense adopted by Harvey. A scientist hopes to be able to link together a number of inductive laws which will include the relationships and association between the established laws. From this material he hopes to formulate general and overriding laws. The weakness of the inductive method is that the processes of ordering and structuring data are not independent of the theory which is ultimately constructed. The *a priori* establishment of a system of classification is an essentially similar operation to the setting up of an *a priori* theory.

Ritter used the inductive method as a framework for his presentation of data and as a means to arrive at some simple empirical generalisation. From this material he hoped to gather evidence about the overriding principles in God's plan for mankind — which he regarded as the underlying purpose of development on Earth. Such a teleological philosophy cannot be tested empirically and therefore does not qualify as scientific explanation. It does, however, have the characteristics of a paradigm as defined above.

Harvey (1969, p438) considered that the teleological framework of explanation is possible without such a metaphysical assertion. A *teleological explanation* is generally taken to mean that a phenomenon is explained in relation to the purpose it is believed to serve. A *mechanical* or *causal explanation*, on the other hand, relies on pre-existing causes to explain the

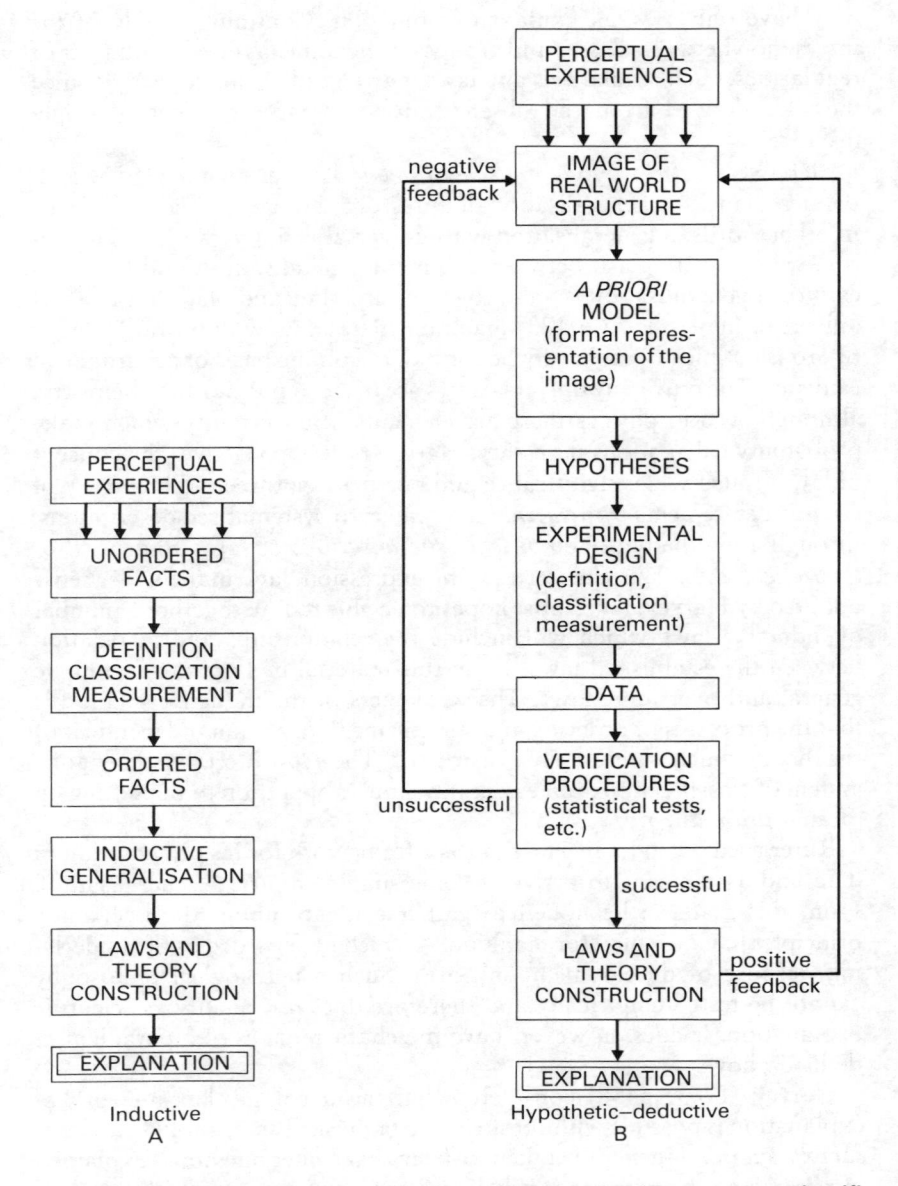

Figure 4. Inductive (Baconian) and hypothetic–deductive routes to scientific explanation. (From Harvey 1969)

observed phenomena. Holistic syntheses and concepts of organic relationships are strongly related to teleological explanatory models.

Ritter combined a teleological approach with a simple deterministic
model in which everything that happens is directed to purposes whose
controlling conditions are laid down by God. Later on, however, concepts
of organic relationships were widely used as alternatives to the deterministic and mechanical explanations of Ratzel, Semple and others. The possibilist argument was that most events are not governed by the operation
of mechanical laws, but happen because men choose specific methods and
actions to achieve their predetermined goals.

Because of Ritter's teleological outlook, it is difficult to identify the
professionalism phase in geography. The first apparently active school of
geography did not lead the subject into the first phase of a paradigm. The
development of Darwinism led to the rejection of ideas which might have
developed into a paradigm and a discussion actually started as to whether
geography as a whole could be regarded as a science or not.

It should be emphasised that Darwinism did not represent a complete
break with the major ideas upon which geography had been founded. The
study of development over time continued to be regarded as very important and, although it is doubtful whether Darwin supported it or not, a
deterministic explanatory framework was retained. There was an inevitability in the development of nature and society; the determinists argued
that natural conditions determined cultural development. The conflict
with Ritter's ideas was not over the deterministic explanatory model itself,
but about the forces that shaped development. After Darwin, scientists
were looking for the controlling laws of nature (and the materially conditioned social laws) and to a considerable extent adopted a nomothetic or
law-making approach.

This development was especially characteristic of the natural sciences,
which were making major advances. Inductive arguments were increasingly replaced by _hypothetic–deductive methods_ (figure 4B). Research
workers, starting from inductive arrangements of their observations or
from intuitive insights, tried to devise for themselves *a priori* models of the
structure of reality. These were used to postulate a set of hypotheses which
could be confirmed, corroborated or rejected by testing empirical data
through experiment. A large number of confirmations led to a *verification*
of the hypothesis, which was then, for the time being, established as a law.
This law stands until its eventual rejection as a result of later research. No
proof of the absolute truth of a law can ever be produced for definite
verification is virtually impossible.

Karl Popper has pointed out that the truth of a law does not depend on the number of times it is confirmed experimentally; it is easy enough to find empirical support for almost any theory. The criteria for its scientific validity are not the confirmatory evidence, but that those circumstances which may lead to the rejection of the theory are identified. It follows that a theory is scientific if it is possible to *falsify*. Kuhn criticises Popper for believing that a theory will be abandoned as soon as evidence is found which does not fit the theory. Kuhn maintains that all theories will eventually be confronted with some data which do not fit. A fundamental theory is not rejected if individual research data do not fit it. It is only rejected when a new theory is put forward which is *believed to be* better (Johansson 1973).

Another problem is that deductive analyses often contain inductive elements. Despite these weaknesses, it has been difficult to find a method which is more tenable from a logical standpoint than the hypothetic–deductive system.

The hypothetic–deductive method has been recognised as the characteristic method of the natural scientist, and widely regarded as the only scientific method. It has been most thoroughly developed in physics. In biology and geology the hypothetic–deductive method is less strongly developed than in chemistry and physics because of the type of the questions which must be answered and the nature of the empirical data studied in these subjects. It is characteristic of the phenomena studied in physics that objectives are quantifiable. The phenomena of theoretical physics are, however, much more 'abstract' than those studied by biologists and geologists. As neutrons and atomic nuclei cannot be directly observed, measurement has provided a theoretical supposition of their existence. Theoretical physics operates in an abstract milieu, seeking unity and association through mathematical hypotheses and postulates. Physics has been able to develop into a model-building theoretical science precisely because it works with abstract and quantifiable phenomena and not because its phenomena are concrete and quantifiable. Geology and the systematic sections of biology study concrete phenomena which have a known situation in time and space. They do not manifest themselves directly as quantities and theory plays a less important role in studying them. The same can be said about the phenomena studied in history, geography and the social sciences. It is significant, however, that psychology and the social sciences which study not so clearly identifiable social phenomena display stronger tendencies to theorising.

An interesting aspect of the development of geography during the latter half of the nineteenth century is the way in which research workers developed the subject as a nomothetic science to a greater extent than might have been expected. The hypothetic–deductive method was, however, not applied in a strict sense; we may agree with Minshull (1970, p81) that determinists as well as geomorphologists stated the generalisations first and then gave a few highly selected examples as proof. Geographers could not test hypotheses by verification procedures involving a number of repeated experiments, as could physicists. Statistical tests that might play the same role as experiments were not then sufficiently developed to cope with the complex geographical material.

Using Kuhn's terminology, geomorphology and determinism represented the first paradigm phase in geography. This paradigm was effective for geomorphology — it lasted for a good half-century and advanced the scientific reputation of the whole subject through the accumulation of scientific information until alternative explanations began to be put forward. While geomorphology expanded through the accumulation of relevant knowledge, the other branches of geography experienced a series of crisis phases. For this reason we can leave geomorphology aside and concentrate on developments in human geography.

Determinism had a short life as the dominating paradigm of scientific method in human geography. It was challenged by possibilists and by the French school of regional geographers who stressed that man has free will and participates in the development of each landscape in a unique historical process. There was also an attack on the methodological front where geographers trained themselves to concentrate on the study of the unique single region. This automatically limited the development of theory (as normally understood in science) and made the hypothetic–deductive method redundant. A corresponding rejection of the inductive method, which also attempts to advance to inductive laws, might have been expected. The appropriate method would be to try to *understand* a society and its habitat through field study of the ways of life and attitudes of mind of the inhabitants of the area concerned. This method, in the form of *participating observation*, characterises the work of many social anthropologists today.

In practice, however, possibilists and regional geographers were somewhat freer in their choice of methods. Fieldwork was, however, regarded as of the utmost importance. Vidal de la Blache based his *Tableau de la Géographie de la France* (1903) on studies in each département. Albert Demangeon walked every lane in Picardy before publishing his regional monograph on that *pays* in 1905. Such fieldwork was not, however, sufficient for geographers who wished to work on a macro scale and to collect data from large regions. They needed material from written sources and from official statistics. Their handling of these data, their organisation, classification and analysis, approached the inductive method very closely. They also sought general causal relationships but were rather unwilling to identify them as 'laws.'

Although possibilists reacted against the simple explanatory models of the determinists, they developed further many ideas derived from Darwinism. They took over Darwin's ideas about struggle and selection although they also held that chance and human will played an important role in development. While possibilism and the regional geographical school established a new paradigm, this new paradigm did not immediately defeat its predecessor. Partly because of the strong position of geomorphology and physical geography, the deterministic explanatory model survived side by side with possibilism and, even within the circle dominated by the new paradigm, the puzzle-solving phase was short-lived.

For a long time, however, geographers continued to stress the central position of regional geography in the subject. Georges Chabot declared in 1950 that 'Regional geography is the centre around which everything converges.' It is however fairly obvious that the greatest advances in geographical research during the last 50 years have taken place within systematic geography. In geomorphology, plant geography, economic geography, settlement geography, population geography and in many other branches, a range of new theories and methods have been developed and research has increased strongly. During the inter-war period, morphological and functional regional geography, landscape ecology and landscape morphology were subdivided into systematic branches. Many specialised studies were made of urban morphology, and research into the morphology of rural settlements was separated from general studies of agrarian cultural landscapes. Morphological research became less evident during the post-war period but there was more interest in specialised functional studies of urban and rural communities. Regional geography, on the other hand, did not make similar progress.

Regional geography had flourished in countries like France, where geography was closely associated with history in school and university teaching and where the educational system fostered a national self-image of sturdy peasantry and cultured townsfolk. Regional studies were also important to the academic leaders of emergent nations in central and eastern Europe who were seeking to establish and preserve the uniqueness of their national heritage, not just through the native language, but also by studying a whole range of traditional relationships to their land which had survived centuries of foreign domination. In Finland, for example, the first editions of a thematic national atlas appeared as early as 1899 and 1910 (when Finland was a Grand Duchy within the Russian Empire) and played an important role in the establishment of national self-consciousness.

While the peace settlement of 1919–21 created many new nation states in Europe, arguments over their boundaries between the 'winners' and 'losers' of World War I continued to draw extensively on local historical and geographical relationships. In 1945 the division of Europe into American and Russian spheres of influence took the geographical re-alignment of nationalities and nation states out of the range of practical politics. The economic revival of western and northern Europe, which has developed through prosperity to affluence since 1945, has been associated with the growth of essentially similar urban industries and services, organ-ised and controlled at national or even international level. 'Regions' have come to be defined in strictly economic terms: 'regional policies' are devised to help areas which lag behind the national norm of economic growth. The communist parties which came to power in Russia (in 1917), in eastern and central Europe, and in China (after 1945) have also, in their different ways, been more concerned with changing society than in pre-serving (and therefore studying) traditional local characteristics. The study of relations between the indigenous inhabitants and their territories has had even less relevance in North America, southern Africa and Australasia, where the destiny of the European settlers, as they saw it, was to occupy the country and to reorganise it in a wholly new and much more productive way.

Another reason why the regional geographical method has more or less disappeared is found in the basic philosophy of the subject held by Hettner and Hartshorne. While both regarded the regional geographical synthesis as central to geography, they discouraged historical methods of analysis, basing themselves on Kant's view of geography as a chorological science. Hartshorne was strongly criticised, by Carl Sauer among others who, within a year of the publication of *The Nature of Geography* in 1940, said 'Hartshorne . . . directs his dialectics against historical geography, giving it tolerance only at the outer fringes of the subject . . . Perhaps in future years the period from Barrows' *Geography as Human Ecology* (1923) to Harts-horne's latest resumé will be remembered as that of the Great Retreat.' (Sauer 1963, p352).

The advantages of historical explanatory models have been appreciated especially by historical geographers and geomorphologists. Accordingly, historical explanations have been used by certain schools of thought within geography up to the present time.

The morphological and functional approach in regional geography seems also to have restricted the presentation of regional syntheses. A contributory reason for this is the growth of data and of scientific activity, both of which demand that research workers should specialise in order to make progress. Individual workers found it difficult to develop new insights when they had to study the total interplay of factors within a region. Instead, they illustrated conditions within a region through studies of certain key factors. For example, the study of population and settlement conditions were regarded as critical to the understanding of regional problems.

The concept of the subject developed by Kant, Hettner and Hartshorne was, however, adopted by a large majority of human geographers from the 1930s until the 1960s. This, if anything, could be regarded as a paradigm. The disadvantage of their approach was that it did not lead to a universally accepted method of chorological regional description, nor did it give clear instructions as to how puzzle-solving was to develop.

Kuhn's model fits the development of geographical science only superficially. As we have followed the development of the subject, we have seen how new paradigms have, to some extent, included ideas from the older paradigms. The paradigm concept therefore loses some of its clarity and value as a guide for research until, in the end, more and more people define geography as what geographers do. Despite the impressions we may get from simplified accounts (for instance Wrigley 1965), a closer look at the history of geography reveals that complete revolutions have not taken place; paradigms, or what may be more appropriately termed schools of thought, continue to exist side by side.

An Idiographic or Nomothetic Science?

Another reason for paradigm shifts being more apparent than real is that each new generation of workers, or each individual trying to change the scientific tradition of his discipline, will tend to ascribe a more fundamental significance to their own findings and ideas than they really have. A characteristic oversimplification of the views held by the immediately previous generation, or, rather, by the leading personalities of the current tradition, is demonstrated a number of times in the history of geography.

A rather good example is the vigorous criticism of Hartshorne presented by Fred Schaefer (1953) in *Exceptionalism in Geography*. Schaefer attacked the 'exceptionalist' view of the Kant–Hettner–Hartshorne tradition; the view that geography is quite different from all other sciences, methodologically unique because it studies unique phenomena (regions), and therefore is an idiographic rather than a nomothetic discipline.

'Hartshorne, like all vigorous thinkers, is quite consistent. With respect to uniqueness he says that "While this margin is present in every field of science, to greater or lesser extent, the degree to which phenomena are unique is not only greater in geography than in many other sciences, but the unique is of very first practical importance." Hence generalizations in the form of laws are useless, if not impossible, and any prediction in geography is of insignificant value. For Kant geography is description; for Hartshorne it is "naive science" or, if we accept his meaning of science, naive description' (Schaefer 1953, p239).

Schaefer maintained that objects in geography are not more unique than objects in other disciplines and that a science searches for laws. Having eliminated some of the arguments against the concept of a rigorous scientific geography, Schaefer sought to set down the kinds of laws which geographers ought to seek, and also urged them to study systematic rather than regional geography.

Schaefer's argument on laws in geography was of importance in changing the direction of geographic research, at least in the United States. 'In the 1950s,' maintained Leonard Guelke (1977a, p383), 'geographers were given a choice between describing the unique or seeking scientific laws. The former alternative was, not surprisingly, unacceptable and there was a basic shift in the discipline away from any consideration of the unique.' Ronald Johnston (1978, p194), however, maintained that the new paradigm was introduced gradually, entering by the side door, if not by the back. The first presentation of its philosophical and methodological entirety, *Explanation in Geography* by David Harvey was only published in 1969.

It is rather interesting to note that Harvey elaborated the theme of exceptionalism in much the same way as Schaefer, and seemed to disregard, more or less, Hartshorne's (1955) very strong counter-attack on Schaefer in which he maintained: 'The title and organisation of the critique lead the reader to follow the theme of an apparent major issue, "exceptionalism," which proves to be non-existent. Several of the subordinate issues likewise are found to be unreal' (Hartshorne 1955, p242). Hartshorne admitted to having used the words idiographic and nomothetic, but rejected the idea that different sciences can be distinguished as being either idiographic or nomothetic, regarding these two aspects of scientific knowledged as being present in all branches of knowledge (*op. cit.* p231). As early as 1925 Sauer suggested that although geographers had earlier been devoted to descriptions of unique places as such, they had also been trying to formulate generalisations and empirical laws.

Both Hettner and Hartshorne made a distinction between systematic geography, which seeks to formulate empirical generalisations or laws, and the study of the unique in regional geography, whereby generalisations are tested so that subsequent theories may be improved. Hartshorne (1959, p121) suggests that geographical studies show 'a gradational range along a continuum from those which analyse the most elementary complexes in areal variation over the world, to those which analyse the most complex integrations in areal variation within small areas.' James (1972, p468) emphasises that there is no such thing as a 'real region.' The region exists only as an intellectual concept which is useful for a particular purpose. Harvey and others have read a much more metaphysical significance into the concepts 'unique' and 'region' than the geographers who were practising between the wars intended. It is also true that incorrect quotations from Hettner and Hartshorne have gained an amazingly wide acceptance. 'It is discouraging to find some writers who continue to accuse Hettner and his followers of defining geography as essentially idiographic, thereby obscuring the underlying continuity of geographic thought' (James 1972, p228). James thus maintains that what we call the quantitative 'revolution' did not represent such a major change in direction as many think. Traditional geography included some of the bases for quantification; rigorous measurements and the testing of hypotheses were normal in geomorphology by the 1940s. Possibilism is linked to probabilism when we consider that men, when changing the face of the Earth, are much more likely to act in some ways than in others. Probability statistics are relevant to such situations.

It can hardly be denied, however, that a significant change in the type of research undertaken by geographers took place in the 1950s and 1960s. The inter-war generation of geographers had been sceptical of the formulation of general and theoretical laws, partly as a reaction against the crudities of environmental determinism. It was only after World War II that such theoretical issues as the study of diffusion models and location theory, and also the search for geometrical models to explain geographical patterns, came to occupy a dominating position in the subject.

The Quantitative Revolution

Location theory, as taught today, originates from economic theory. The classic location theories, including Johan Heinrich von Thünen's work on the use of areas in agriculture (1826) and Alfred Weber's study of industrial location (1909), are economic theories. Later economists, including Ohlin, Hoover, Lösch and Isard, have developed our understanding of the areal and regional aspects of economic activity.

Walter Christaller (1893–1969) was the first geographer to make a major contribution to location theory with his famous thesis *Die Zentralen Orte in Süddeutschland* (1933), translated by Baskin as *Central Places in Southern Germany* (1966). Christaller, who had studied economics under Weber, declared in 1968 that his work was inspired by economic theory. His supervisor when working on *Die Zentralen Orte* was Robert Gradmann, a geographer who had himself made an outstanding regional study of southern Germany (1931) which, however, closely followed the current idiographic tradition in German *Länderkunde*. Although Christaller's thesis was accepted, his work was not acclaimed during the 1930s and when Carl Troll (1947) wrote a review of what had been going on in German geography between the wars, he did not even mention him. In Kuhn's terminology, Christaller's attempt to explain the pattern and hierarchy of central places by a general theoretical model was not acceptable within the reigning paradigm. Christaller never held an official teaching position in geography.

Eventually Christaller gained a following, notably in North America and Sweden. Edward Ullmann was one of the first American geographers to draw attention to Christaller's work (Ullmann 1941). American economists and urban sociologists had developed theoretical models of urban structures and of cities as central places rather early, and geographers soon followed these leads (Harris and Ullmann 1945). This development and the philosophical discussion which followed the debate between Schaefer and Hartshorne during the 1950s led to an early acceptance of theory-building and modelling. In this respect a very well balanced and clear account of geography as a fundamental research discipline by Edward A Ackerman (1958) encouraged students to concentrate their attention on systematic geography, cultural processes and quantification. A range of different statistical methods was gradually brought into use in several systematic branches of geography enabling the development of more refined theories and models.

This acceleration of theoretical work was especially marked in institutions led by geographers who had studied the natural sciences, especially physics and statistics, and/or where there were good contacts with developments in theoretical economic literature. The frontier between economics and geography became very productive in new ideas and techniques during the 1950s at several American universities (Garrison 1959–60). A seminar for PhD students in the use of mathematical statistics conducted by William L Garrison at the University of Washington, Seattle, from 1955 onwards was of particular significance. Garrison and his co-workers were mainly interested in urban and economic geography, into which they introduced location theory based on concepts from economics with associated mathematical methods and statistical procedures (Garrison 1959–1960). Many of the students from Seattle became leaders of the 'new' geography in the United States during the 1960s, including Brian J L Berry, William Bunge and Richard Morrill. Both Berry and Garrison later moved to work in the Chicago area where Berry had much to do with the development of the long established geography department at the University of Chicago into a leading centre of theoretical geography. There was a simultaneous development of quantitative geography at the Universities of Iowa (where Schaefer had taught until his death in 1953) and Wisconsin.

It is possible, as Johnston (1979) maintains, to recognise four schools of quantitative geography in the United States. Three were developed in the departments of geography in the Universities of Washington, Wisconsin and Iowa, with Washington as the most prominent centre of innovation. The fourth, 'social physics' school, developed independently, drawing its inspiration from physics rather than economics. Its leaders were John Q Stewart, an astronomer at Princeton University, and William Warntz, a graduate in geography from the University of Pennsylvania (who was later employed as a research associate by the American Geographical Society). In the 1920s American sociologists had postulated that the movement of persons between two urban centres would be proportional to the product of their populations and inversely proportional to the square of the distance between them, but it was Stewart who pointed out the *isomorphic* (equal form or structure) relationship between this empirical generalisation and Newton's law of gravitation. Thereafter this concept became known as the *gravity model*. Stewart's ideas about isomorphic relations between social behaviour and the laws of physics were introduced to geographers by a paper in the *Geographical Review* as early as 1947. Here Stewart stated that human beings 'obey mathematical rules resembling in a general way some of the primitive "laws" of physics' (Stewart 1947, p485). Warntz, working with Stewart, also borrowed analogy models from physics in his studies of population potentials (Warntz 1959, 1964). He suggested that the mathematics of population potential is the same as that which describes a gravitational field, a magnetic potential field and an electrostatic potential field (James 1972, p517).

The work of Christaller, August Lösch and others was introduced into Sweden by Edgar Kant, an Estonian geographer who had tested their theories in his homeland before taking refuge in Lund after World War II (Kant 1946, 1951). His research assistant in 1945–6 was Torsten Hägerstrand, a brilliant young geographer, who was also working on migration processes. Through his contacts with the Swedish ethnologist Sigfrid Svensson, who had made a number of studies of the relations between innovation and tradition in rural areas using the currently accepted methodology, Hägerstrand became interested in the possibilities of investigating the process of innovation with the aid of mathematical and statistical methods. In focusing on the *process* Hägerstrand made a clear break with the current regional tradition. His dissertation *Innovationsförloppet ur korologisk synpunkt* (1953, later translated by Pred (1967) as *Innovation Diffusion as a Spatial Process*), examined the diffusion (or spread) of several innovations among the population of a part of central Sweden. Some of these innovations concerned agricultural practices, such as bovine tuberculosis control and pasture improvement, and others were more general, such as car ownership. Hägerstrand's work is less important for its empirical findings than for its general analysis of the diffusion process. He stated himself that although the material used to throw light on the process relates to a single area, this should be regarded as a regrettable necessity, rather than a methodological subtlety (Hägerstrand 1953, 1967, p1). With the aid of the so-called 'Monte Carlo simulation', which involves the use of random samples from a known probability distribution, he was able to construct a general *stochastic* model of the process of diffusion. Stochastic literally means at random; stochastic or probability models are based on mathematical probability theory and build random variables into their structure. Models may be classified as either stochastic or *deterministic*. In deterministic models the development of some system in time and space can be completely predicted, provided that a set of initial conditions and relationships is known.

The stochastic Hägerstrand model enabled the spread of innovation to be simulated and later tested against empirical study. It was demonstrated that the form of distribution at one stage in the process would influence distribution forms at subsequent stages. Such a model *could* therefore be of use to planners in support of future innovations which they wished to bring about.

The department of geography at Lund University soon became renowned as a centre of theoretical geography, attracting scholars from many countries. Almost from the beginning there were contacts between Lund and Seattle. Hägerstrand taught in Seattle in 1959 and Morrill studied with him in Lund, where his work on migration and the growth of urban settlement (Morrill 1965) was presented.

In the years that followed, Hägerstrand's technical and statistical procedures attracted more attention than his analyses (1953, p169 ff) of individual fields of information and their change through time. He regarded the study of such information fields as basic to the deeper understanding of the processes of diffusion. During the 1960s Hägerstrand went on to make detailed studies of individual behaviour, using three-dimensional models to portray the movement of individuals in time and space. An important feature of *time–space geography* is that time and space are both regarded as resources which constrain activity. Individuals have different possibilities of movement in space, conditioned by their economic status and technical possessions, but time imposes limitations on everyone. Subsequent studies in time–space geography, which have been carried out actively at Lund and elsewhere throughout the 1970s and are summarised in Carlstein (1978), have shed much new light on patterns of diffusion and other geographical aspects of human behaviour.

In 1963 a Canadian geographer Ian Burton, arguing that the quantitative revolution was over and had been for some time, cited the rate at which schools of geography in North America were adding courses in quantitative methods to their requirements for graduate degrees. It must be stated, however, that the theoretical development within the subject was not felt as a revolution by most geographers, and many 'revolutionaries' were at pains to emphasise continuity in the ultimate objectives of human geography. The use of statistics for the making of relatively precise statements was generally accepted, although the related use of mathematics in modelling received much less attention (Johnston 1978).

Most research workers regarded advanced statistical methods as being useful in some branches of the discipline: other branches, notably historical and cultural geography, have felt less need for new techniques. Leonard Guelke (1977b, p3) has claimed that 'To an extent that is not widely recognised, the move to quantification took place within the basic framework of geography put forward by Hartshorne in *The Nature of Geography* . . . [Hartshorne's] widely diffused ideas were essentially well disposed to the adoption of the new methods.' In many geography departments the works of both Hartshorne and Garrison were on the students' reading lists, but philosophical and methodological differences between them were not an issue in teaching up to the mid-1960s; Schaefer's viewpoints had been forgotten.

Johnston (1979, p62) points out that the leaders of the quantitative school did not study the philosophy which they were adopting very deeply, apart from references to the works of Gustav Bergmann, a philosopher and close friend of Schaefer who had actually read the galley proofs for Schaefer's paper on 'exceptionalism' (Schaefer 1953) in some of the papers of the Iowa group and William Bunge's thesis *Theoretical Geography* (1962, new edition 1966). Bunge, who had worked at Iowa for a short period, extended the arguments of Schaefer, to the effect that geography is the science of spatial relations and interrelations, geometry is the mathematics of space, and so geometry is the language of geography. The *chorological* viewpoint, emphasising nature and interrelationships between specific places or regions was rejected in favour of a geography based on *spatial analysis* which stressed the geometric arrangement and the patterns of phenomena. "The study of the where of things, their spatial distribution, which the regional, chorological view considered as a deviation, was presented as the core of the geographic enterprise." (Sack, 1974 p444).

The major advances towards a unifying methodological and philosophical basis for the quantitative school were made in the 1960s by British geographers, notably Peter Haggett, Richard Chorley and David Harvey. *Locational Analysis in Human Geography* by Peter Haggett was published in 1965. The importance of this book lay in its overview of much new theoretical work in the subject. Haggett (1965, pp14–15) used the diagram reproduced in figure 5 to illustrate the argument that there are really three traditional subject associations of geography: with the Earth sciences (geology and biology), with the social sciences and with the geometrical sciences.

The geometrical tradition, the ancient basis of the subject, is now probably the weakest of the three, he maintained. 'Much of the most exciting geographical work in the 1960s is emerging from applications of higher order geometrics . . . Geometry not only offers a chance of welding aspects of human and physical geography into a new working partnership, but revives the central role of cartography in relation to the two' (Haggett 1965, pp15–16).

Haggett's book led to a fundamental debate within the subject. The

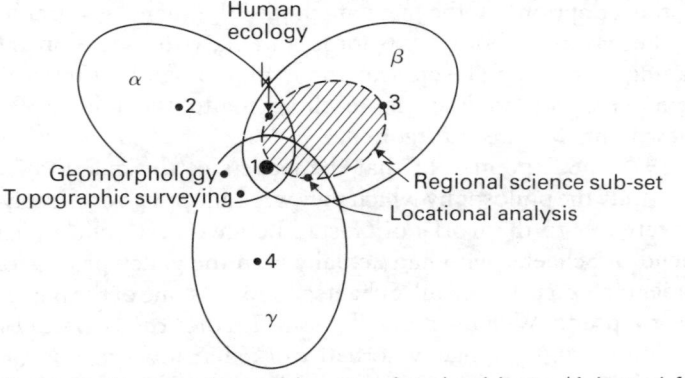

Figure 5. Geography and its associated subjects. (Adapted from Haggett 1965)

arguments presented by Kuhn (1962, 1970) on paradigm shifts within the world of science were applied in the debate. Thus Chorley and Haggett (1967, p39) stated that they had looked at the traditional paradigmatic model of geography and had found that it was largely classificatory and under severe stress. They suggested that geography should adopt an alternative *model-based* paradigm, and so made it clear that the new development within the subject not only represented a wider range of methods, but demanded a fundamental paradigm shift. Each geographer was given the choice between the traditional and the new model-based paradigm. Model building was set up as the aim of geographical investigation, a task to be performed with the aid of *quantitative methods* and the use of computers to handle data. A model was defined as an idealised or simplified representation of reality which seeks to illumine particular characteristics. The concept is a wide one — for Chorley and Haggett (1967) a model was either a theory or a law or an hypothesis or a structured idea.

The rapid development of model building and the use of quantitative techniques could not have taken place without computers, but the computers did not determine the development of either model building or quantitative methods. 'Model building preceded the invention of the computer in many sciences, but in a discipline like geography which handles such large quantities of data it would hardly have been possible to develop operational models worthy of the name without computers' (Aase 1970, p23). This technological development had given the subject new possibilities which researchers had no hesitation in exploring.

The quantitative 'revolution' did not take place without opposition. Dudley Stamp (1966, p18) preferred to term the quantitative 'revolution' a 'civil war,' and noted that quantification had many points in common with a political ideology; it was more or less a religion to its followers, 'its golden calf is the computer.' Jan O M Broek (1965, p21) stated that 'there are more things between heaven and earth than can safely be entrusted with a computer.' Even Ackerman, one of the advocates of quantification, warned (1963, p432) that 'the danger of dead end and nonsense is not removed by "hardware" and symbolic logic.'

Stamp (1966, p19) pointed out that there are many fields of enquiry in which quantification may stultify rather than aid progress, because there will be a temptation to discard information which cannot be punched on a card or fed onto a magnetic tape; there is also a danger that ethical and aesthetic values will be ignored. Broek (1965, p79) voiced the opinion that the search for general laws, at a high level of abstraction, 'goes against the grain of geography because it removes place and time from our discipline'. Roger Minshull (1970, p56) observed that the landscape was becoming a nuisance to some geographers, that many of the models will only apply to a flat, featureless surface, and warned that there was a real danger that these ideal generalisations about spatial relationships could be mistaken for statements about reality itself.

The hypothetic–deductive method implies that a hypothesis or model is made first. This is an excellent way of avoiding collecting facts for the sake of collecting facts, but Minshull (1970, p128) suspected there would be an overriding temptation not to test and destroy one's beautiful hypothesis or model but to prove it in a subjective way. In the knowledge that there are subjective elements in even the apparently objective sections of a verification procedure, such as classification (Johnston 1968), this possibility is clearly present. Fred Lukermann (1958) reacted especially to attempts by the social physics school to establish analogies with physics, maintaining that hypotheses which are derived by analogy cannot be tested: falsification is impossible. Robert Sack, a former associate of Lukermann at the University of Minnesota, criticised the view put forward by Bunge (1962) and Haggett (1965) that geography is a spatial science and that geometry is the language of geography in a series of papers in the early 1970s. Sack (1972) maintained that space, time and matter cannot be separated analytically in a science which is concerned with providing explanations. The geographical landscape is continuously changing. The processes which have left historical relics and which are creating new inroads all the time must be taken into account as important explanatory factors. The laws of geometry are, however, static — they have no reference to time. The laws of geometry are sufficient to explain and predict geometries, so that if geography aimed only at analysis of points and lines on maps geometry could be sufficient as our language. But, 'We do not accept description of changes of its shape as an explanation of the growth of a city . . . Geometry alone, then, cannot answer geographic questions' (Sack 1972, p72).

One clear deficiency of quantitative generalisation, as Broek reminds us (Broek 1965, p79), is that 'since massive quantitative data on human behaviour are only available for the advanced countries, and then only for at best a century, the theorists tend to construct their models from facts of the "here and now," virtually ignoring former times and other cultures. The procedure becomes invidious when one projects the model derived from one's own surroundings over the whole world as a universal truth and measures different situations in other countries as "deviations" from the "ideal" construct.'

There is certainly a danger that models based on research within the western cultural experience may be elevated into general truths. Brian Berry (1973b) came to the conclusion that a universal urban geography does not exist, and that urbanisation cannot be dealt with as a universal process: 'we are dealing with several fundamentally different processes that have arisen out of differences in culture and time' (Berry 1973b, pXII). He divided the world into four universes: (1) North America and Australia, with their free market economies; (2) Western Europe, with its planned welfare economy; (3) the Third World, with its economy split between a traditional and modern sector, and (4) the Socialist countries, with their rigidly planned economies. Each of these has its own urban geography, which again will change through time.

Haggett, Cliff and Frey (1977, p24) also noted that 'the Russian translation of the first edition of this book (Haggett 1965) made clear how heavily the locational explanations were rooted in the classical economics of the capitalist world. Inevitably, the lopsidedness of the book will appeal to certain readers and condemn it to others.'

Whereas the adherents of the quantitative school could admit a certain lopsidedness in their approach by the late 1970s, their approach and argumentation had been far more orthodox at the end of the 1960s when it was thought that a definite choice between paradigms had to be made.

Absolute and Relative Space

Harvey (1969) argued that Kant's concept of geography as a chorological science was not tenable because it built on the assumption of *absolute space*. The concept of absolute space is tied to Euclidean geometry which is based on five axiomatic statements; the parallel postulate (which states that, given a straight line and a point outside it, there is only one straight line through that point which is parallel to the given line) is one of these. In the Euclidean system a straight line is defined as the path of the shortest distance between two points. In the nineteenth century, however, mathematicians including Carl Gauss, Nicolai Lobatschevsky and Bernhard Riemann showed how to construct a non-Euclidean geometry. Gauss, who gave his name to a celebrated orthomorphic map projection, observed that Euclidean geometry is relevant to two-dimensional space, but it is also possible to regard space as spherical, in which case the shortest line between two points is the arc of a great circle, and the parallel postulate does not apply. Riemann showed that hyperbolic, Euclidean and elliptic geometries were special cases of what came to be known as the geometry of 'Riemannian space.' The essential point of this approach is that the type of geometry required is a consequence of the rules adopted for making spatial measurements (Harvey 1969, p201). It is also possible, given Riemann's general theory, to extend space to more than three dimensions; n-dimensional space can be discussed. It was left to Albert Einstein to provide an application of Riemannian general theory.

As science has accepted Einstein's theory of relativity, it has rejected the concept of absolute space. It is ironic, observes Harvey (1969, p209), that such influential geographers as Hettner and Hartshorne took their guidance from Kant, rather than from Gauss, who also directly made major contributions to the science of geography in his work with map projections. As a consequence the main current of philosophical opinion within geography in the first half of the twentieth century was based upon concepts which other scientists had already rejected as untenable. We cannot identify any point which objectively represents 'now': both time and space are relative concepts. Nothing in the world of physics may be characterised as pure chorology or pure chronology; everything is process. Abler et al (1972, p72) express a similar viewpoint: 'The shift to a relative spatial context is still in progress and is probably the most fundamental change in the history of geography as it opens an almost infinite number of new worlds to explore and map.'

When it comes to practical methods, however, it is quite clear that absolute location on an isotropic surface is the most common way of viewing space. This is what ordinary maps portray, regardless of the problems created by the transformation through map projections of the spherical globe to a plane map. Especially when mapping smaller areas, say a part of England, where a plane surface is a good approximation of

the globe, the distortion created by projections is negligible. This is generally the scale at which geographers work. In a practical sense Euclidean
space is rather useful, but it may be argued that 'faced by the seductive
utility of Euclidean space we have allowed an interest in maps to become
an obsession' (Forer 1978, p233).

Two new basic lines of geographical research each give a central importance to concepts of *relative space*. Recent work in behavioural geography
(see p71 for a discussion of the concept) minimises the significance of
space as it is mapped objectively, but emphasises instead the importance of
the perception of space. Spaces become relative because they are related to
the perceptions of individuals. In location theory the builders of spatial
models are also trying to use measures of relative rather than absolute
location.

The main point is that most concepts of accessibility or isolation refer to
distance measured in a special way, usually in terms of cost distance, time
distance or mileage through a transportation network, and these distances
are measured from special nodes or axes. One important aspect is that
geographical features like settlement patterns, land-use, diffusion processes, etc, show a location and dynamics which to a large extent are due to
their relative positions in space.

Pip Forer (1978, p235) observes that since distances in time, cost or even
network mileage are partly artefacts of socioeconomic demands and technological progress; these types of spaces are naturally dynamic and truly
relative. This leads him to the definition of *plastic space*, a space that is
continuously changing its size and form. An illustration is given with his
own time–space map of New Zealand (figure 6) (Forer 1978, p247).

This discussion leads to the following conclusions: the Kantian view of
space is false and therefore it is not possible, on a philosophical basis, to
define geography as a strictly chorological science and to contrast it with
history, which is defined as the study of all phenomena organised according to the time dimension. On the same grounds we cannot argue that
studies which emphasise process should be regarded as 'not geographical'.
The line of argument proposed by Harvey leads to the conclusion that we
cannot define the content and methods of geography from philosophical
arguments. As Harvey observes (1969, p8), it means that every method is
open for us to use, provided that we can show that its use is reasonable
under the circumstances, but it also means that what may be termed the
Kant–Hettner–Hartshorne paradigm for geography is torn down, without
being replaced with a new, clear-cut paradigm. It is rather difficult to
define a paradigm on a methodological basis alone, especially when all

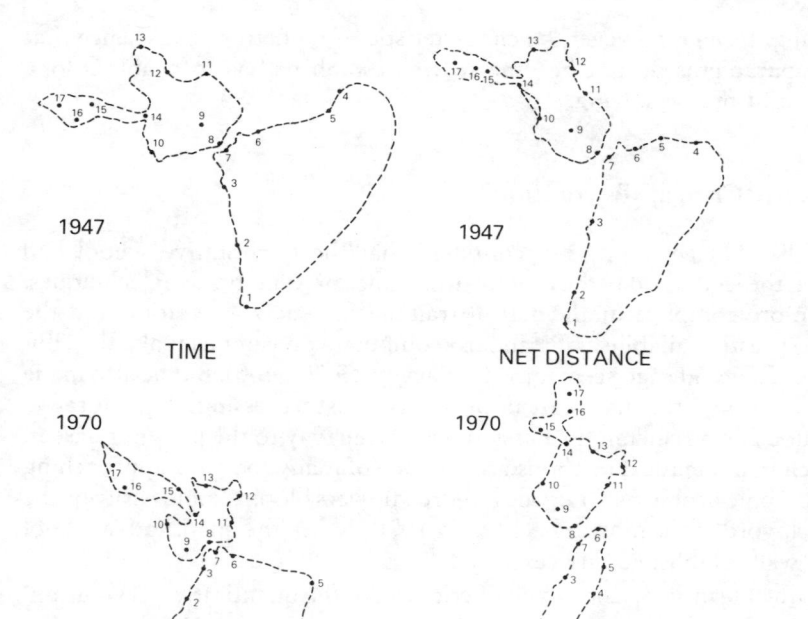

Figure 6. A demonstration of the plasticity of space. The four maps have been constructed from data on the New Zealand airline system and its changes from 1947 to 1970. The two maps on the left show how distance measured in time has changed as the airline network has grown and the speed of travel has increased. The maps on the right show how the net distance travelled has changed with the network. (From Forer 1978)

methods are open for use, provided they are reasonable under the circumstances. The increased use of computers and mathematical–statistical methods hardly represent a new paradigm. It may be argued that Chorley and Haggett (1967) tried to establish a new paradigm with a model-based approach, but they refrained from excluding research with a traditional approach from geography. This meant that the model-based paradigm did not fulfil all the requirements set up by Kuhn in his definition.

The renewed discussion on the basic problems of the subject which followed in the wake of the quantitative 'revolution' may, however, be an appropriate sign of a crisis. Individual research workers felt themselves more or less obliged to take a stand and to clarify their own research situation so there was little opportunity for straightforward puzzle-solving. It may, however, be characteristic of modern social science that new paradigms do not become so well established as to enable a long period of puzzle-solving.

A New 'Critical' Revolution?

In 1972 Haggett appeared confident that the quantitative school had taken the lead. 'Today the general acceptance of [quantitative] techniques, the more complete mathematical training of a new generation and the widespread availability of standard computers on campus make the conflicts of a decade ago seem unreal' (Haggett 1972, p460). But he also made the point that 'the first years of over-enthusiastic pressing of quantitative methods on a reluctant profession have given way to the present phase in which mathematical methods are just one of many tools for approaching geographic problems.' 'Let one hundred flowers bloom' could easily be the watchword of the more relaxed early 1970s, when the quantitative school was well established and respected.

In the meantime, a new type of criticism of the quantitative 'revolution' was developing. In *Directions in Geography* (ed Chorley 1973) a number of geographers who, one way or another, had had some hand in the quantitative innovations, discussed possible directions which the discipline might follow in the future. Many of the contributors suggested quite new directions for research, while others criticised different aspects of the quantitative approach openly. Harvey also became a notable apostate, declaring that 'the quantitative revolution has run its course, and diminishing marginal returns are apparently setting in . . . Our paradigm is not coping well . . . It is ripe for overthrow' (Harvey 1973, pp128–9).

In retrospect it may be said that the 1960s can be characterised as an era of 'hard science', whereas in the 1970s there was much questioning of the law-seeking approach. Guelke (1977a, p385) concluded that 'the idea that a scientific discipline must necessarily have laws of its own is false. A discipline can be scientific if it uses or consumes laws from other areas.' Michael Chisholm (1975, pp123–5) noted that geographical 'laws' in general would not meet the exacting specifications needed to qualify as laws, since they are not generally verifiable. It would be more correct to talk of models and theories rather than laws in geography, a *theory* being defined as an articulated system of ideas or statements held as an explanation (Chisholm 1975, pp123–6).

In his view the essential characteristic of central place theory, and other

theories established by the quantitative school, is their *normative* character; the theoretical construct is not intended to show how the world is
actually organised, but how it should be organised.

Descriptive, or what Chisholm called *positive*, theories seek to account
for observed phenomena, as did, for example, Copernicus' theory of the
motion of the planets around the Sun. The urban rank size relationship is
one of the more famous regularities observed in geography, and may as
such qualify as a positive theory, according to Chisholm (1975, p148). In
positive theories the observations of discrepancies between the predicted
state and actual states of the system may stimulate an advance or changes
in the theory whereas the normative theory is used to create a world that is
'rational.' Normative theories are rather useful in social studies and in
town and regional planning, in which many trained geographers found
career outlets in the 1960s. The many *assignations* in this field, observes
Olof Wärneryd (1977, p29), encouraged a situation in which geography
lost contact with its earlier scientific traditions. As in other fields of social
activity, there was a general belief in economic growth and economic
theories; values embedded in these were not seriously debated.

An example from Sweden, discussed by Gunnar Olsson (1974), may
clarify this point. In the 1960s Swedish geographers were engaged in a
far-reaching reform of administrative and political districts, which was
expressly intended to abolish the spatial elements of social and economic
inequality in the country. The new units were intended to be large enough
to sustain the considerable burden of the welfare state. The methodology
through which this new equality was to be implemented was, however,
through observations of how the majority of people actually interacted in
space, which were then translated into the language of a variant of the
gravity model. 'Unnoticed to spectators and performers, the play was
changed in the middle of the act. The *ought* of justice disappeared into the
wings, invisibly stabbed by the *is* of the methodology. Exit man with his
precise visions, hopes and fears. Enter Thiessen polygons with crude
distance minimisations and cost–benefit ratios' (Olsson 1974, p355).

The arguments presented by Olsson differ somewhat from those of
Chisholm cited above. Olsson talks about methods and theories which
describe what people actually do; such theories are apparently regarded as
normative by Chisholm. In reality there may be little disagreement between them. The quantitative geography of the 1960s made use of theories
from other sciences, notably economics, which were thought to give an
objective description of society and how it functions. Models were constructed which gave an apparent explanation, but were misleading or

directly fallacious in so far as they failed to take account of those underlying explanations of the actual social situation which themselves may be amenable to change. It is possible to argue that such explanatory models tend to support the existing conditions of society. This is particularly true of process models, which encourage us to believe that a trend, once ascertained, will continue to operate in the future. Thus a theory which appears to be positive or based on a realistic description is translated into the purpose of the development through planning and political decision. During this sequence of events the theory becomes normative. In the Swedish example, a planning decision which had the objective of changing a social structure in fact led to the conservation of the structure because of the theories and methods used in the planning process. One conclusion that may be drawn from this is that the distinction between normative and positive theory may be useful in a scholastic sense, but the dichotomy may also be misleading as the same theory can be used both in a normative and a positive way.

Another conclusion, drawn by a number of geographers in the late 1960s, was that physical planning had not been as effective in fostering social change and equality as many people had hoped. For example, many of the land use and transport plans which spread from North America to practically every large city in western and northern Europe, and in which many trained geographers had participated, seemed to have increased the segregation of social classes and to have sharpened differences in mobility between the car-owning and the car-less groups. In transport planning, the interaction pattern of the average family had been used as a guideline. Such 'deviant' travel patterns as those of old people with no access to a car, had not been given much attention. The reason for this had been methodological; quantitative models were built to cope with aggregate and 'hard' data, that is, data which are easily expressed in numbers; 'soft' data, which concern human attitudes and deviations in behaviour, could not easily be handled in such models. But even research workers involved in aggregate studies were bound to wonder about the deviations from the 'normal'. This led to studies of the welfare of special groups of people such as old people in the transport planning example. Perhaps more important was a growing concern for the position of the individual within a mass society.

The students of locational theory also had second thoughts as they came to realise that 'economic man,' that decision-maker who is blessed with perfect predictive ability and knowledge of all cost factors, does not in fact exist. Locational decisions may be made on a rational basis, but this rationality relates 'to the environment as it is perceived by the decision-

```
"TRADITIONAL"            "QUANTITATIVE"          "CRITICAL"

Historical geography ─────────────────────────────────────────────►
Cultural geography ──────────────────────────────────────────────►
                                          Humanistic geography ──►
Landscape studies ──────────────────────────────────────────────►
Regionalisation ──────────(Computer mapping) ───────────────────►
Population and settlement ───────────────────────────────────────►
studies
                        Central place theory ────────────────────►
Economic geography ─────────────────────────────────────────────►
                        Location theory──────────────────────────►
                                          Behavioural geography ──►
                        Social physics───────────────────────────►
                        Innovation studies ──────────────────────►
                                          Time–space geography──►
                        Systems analysis─────────────────────────►
                                          Human ecology ─────────►
                                          Welfare geography──────►
                                          Radical geography──────►

   1950                    1965                   1980

              In current practice ──────────►
```

Figure 7. Schools of human geography shortly after the mid-twentieth century. It must be stressed that this classification should be treated with caution, for the criteria for the classification are not consistent throughout the scheme. While the 'traditional' schools are mainly classified on the basis of their themes, most, but not all, of the 'quantitative' schools are classified in terms of their type of theory. The 'critical' schools are a real mixture, some are classified on the basis of their special methodology, while others owe their position to a particular political approach. Systems analysis is better regarded as a general method which might be useful to a number of schools rather than as a school in itself. The expressions 'traditional,' 'quantitative' and 'critical' are in inverted commas because the use of these terms as labels is highly debatable—as is the grouping of schools under each label.

maker, which may be quite different from either "objective reality" or the world as seen by the researcher' (Johnston 1979, p114). It was thus necessary to derive alternative theories to those based on 'economic man' and to investigate the behaviour and perceptions of the decision-makers. There are a number of different research trends within present-day geography which stem from these observations. Figure 7 gives a simplified classification of those which have emerged since World War II. The 'traditional' and 'quantitative' schools have been dealt with earlier: *systems analysis* and *human ecology* will be dealt with in Chapter 5. Other trends within 'critical' geography will be briefly outlined below, while a discussion of the philosophical basis of 'critical' geography is postponed to the next chapter.

A growing concern for the individual has led some research workers to favour a revitalisation of methods which have already been used by the traditional schools in geography, including those of the French school of regional geography. This trend, which may be seen as a new initiative and as a critique of the quantitative movement by cultural and historical geographers who had not been involved in the model-oriented approach of the 1950s and 1960s, emphasises the need to study unique events rather than the spuriously general. Anne Buttimer (1978, p73) argued along similar lines to Vidal de la Blache, that historical and geographical studies belong together. She stressed the need to understand each region and its inhabitants from the 'inside,' that is, on the basis of the local perspective, and not from the perspective of the researching 'outsider.' Leonard Guelke (1974, p193) advocated an *idealist* approach which 'is a method by which one can rethink the thoughts of those whose actions he seeks to explain.' Guelke considered that geographers should discover what the decision-maker believed, not why he believed it. Thus the human geographer does not need to develop theories, since the relevant theories which led to the action under study already existed in the minds of the decision-makers. Idealism implies one type of *hermeneutic approach*, which will be discussed in more detail later (see pp87 and 96), and *phenomenology* implies another. Whereas idealism accepts that there is a real world outside the individual's consciousness, phenomenologists argue that there is no objective world independent of man's experience; all knowledge proceeds from the world of experience and cannot be independent of that world. One of the best known phenomenologists, Yi-Fu Tuan, who has written a number of inspiring essays and books (1974, 1976, 1977, 1980), has stated (1971) that geography is the mirror of man; to know the world is to know oneself. The study of landscapes is the study of the essence of the societies which mould them. Such study is clearly based in the humanities, rather than in social and physical sciences. 'The model for the regional geographers of humanist leaning is . . . the Victorian novelist who strives to achieve a synthesis of the subjective and the objective' (Tuan 1978, p204). Tuan prefers to use the term *humanistic geography* for such studies, which are regarded here as including both idealism and phenomenology.

Nicholas Entrikin (1976, p616) makes the point that the humanist approach is best understood as a form of criticism. Whereas the 'quantitative' movement was characterised by a great numerical superiority of practitioners over preachers, notes Johnston (1979, p138), 'the phenomenological movement (like the idealist) has been characterized by the converse — much preaching and little practice.'

In contrast to humanistic geography, *behavioural geography* may be seen as a developing criticism from within the 'quantitative' movement, starting from disillusion with theories based upon the concept of 'economic man.' The roots of behavioural geography are, however, much older. In Europe the Finnish geographer Johannes Gabriel Granö and his Estonian student Edgar Kant were attempting a behaviourist approach in the 1920s (Granö, 1929). Even in the United States, the behavioural and quantitative approaches were contemporary developments. The behavioural approach was taken up in the late 1950s and the 1960s by Gilbert White, then at the University of Chicago, and his associates, who made a series of investigations into the human response to natural hazards, guided by theories of decision-making and influenced by methods used in psychology and sociology. It was regarded as more important to map the personal *perception* of the decision-maker than to describe the factual physical and economic conditions of the environment, since the decision-maker would act upon his own perceptions, and not on the environmental factors themselves (White 1973). Julian Wolpert introduced behavioural geography to many human geographers through a paper in 1964 which compared actual with potential labour productivity on farms in central Sweden. He found that the sample farm population did not achieve profit maximisation, nor were its goals solely directed to that objective. The farmers were 'spatial satisfiers' rather than 'economic men.'

One other aspect of behavioural analysis has been the concept of the *mental map* of the environment. Mental mapping has been taken up by a number of workers, among them Rodney White and Peter Gould (1974). A somewhat different approach to behavioural work is found in Allan Pred's (1967, 1969) two-volume work *Behaviour and Location*, in which he tried to present an ambitious alternative to theory-building based on 'economic man.' The *time–space geography* which Hägerstrand established with his associates (see p58) may also be seen as a critique, not so much of the 'quantitative movement' as of important aspects of social science research in general. In simple terms, it provides a method of mapping spatial behaviour and at the same time represents a reorientation of scale, away from aggregate data towards studies of individual behaviour. More important is its introduction of a new economic theory in which time and space are regarded as scarce resources, the allocation of which form the basis of the social realities we study.

A corresponding concern for the individual within mass society is also basic to *welfare geography*, which developed as a special branch in the 1970s. Paul Knox (1975) stated that it was a fundamental objective for geography to map social and spatial variations in the quality of life. The study of such spatial inequalities has been taken up by Bryan Coates and Eric Rawstron (1971), Coates *et al* (1977), Smith (1979) Morrill and Wohlenberg (1971), and a number of others. While some of this work represented the geographer as a 'delver and dovetailer,' a provider of information, other examples, notably *The Geography of Poverty in the United States* (Morrill and Wohlenberg 1971) also proposed both social and spatial policies for changing existing conditions.

Whereas welfare geography works in principle within the framework of the existing economic and social system, *radical geography*, which has become established more recently, calls for both revolutionary theory and revolutionary practice. Its aim is clearly voiced by Harvey in *Social Justice and the City* (1973, p137): 'Our objective is to eliminate ghettos. Therefore, the only valid policy with respect to this objective is to eliminate the conditions which give rise to the truth of the theory. In other words, we wish the von Thünen theory of the urban land market to become *not* true.' Harvey believes that this can only be done if the market economy is eliminated, so the task is the self-conscious construction of a new paradigm for social geographic thought, which may stimulate a political awakening and start a social movement with the ultimate goal of bringing about a social revolution. Geographical science would then become a tool in a Marxist revolution.

To this Berry (1974) commented that Harvey relied too much on economic explanations; in our post-industrial society the control is no longer economic but political, making a Marxist analysis more or less passé. Morrill (1974), however, in reviewing Harvey's *Social Justice and the City* (1973) confesses that he is pulled most of the way by the revolutionary analysis, but that he could not make the final leap that our task is no longer to find truth, but to create and accept a particular truth.

Revolution or Evolution?

Just as Harvey favours revolution in society, so he supports the paradigm concept and Kuhn's theory that science develops through revolutions. 'A quick survey of the history of thought in social sciences shows that revolutions do indeed occur' (Harvey 1973, p122): this is a convenient standpoint for those who wish to revolutionise scientific practice. They can divide people into two categories: those who are not with us are against us.

Harvey (1973, p121) does not, however, regard Kuhn himself as a revolutionary, because of 'his abstraction of scientific knowledge from its materialistic base.' Kuhn provides a dialectic–idealistic interpretation of scientific advancement according to Jan Widberg (1978, p9), who maintains that it is possible to describe the history of scientific thought within a discipline on the basis of four fundamentally different views. These are categorised through two sets of dichotomies:

	Mechanical	Dialectical
Idealistic	I	II
Materialistic	III	IV

A *mechanical* view implies that the development of science is linear, with each new generation continuing from where the old generation stopped. There are no revolutions, only growing specialisation, professionalisation, and the advancement of better methods and understanding. The *dialectical* view is expressed by Kuhn when he maintains that a science develops through contradictions and revolutions with changes of paradigms.

An *idealistic* viewpoint implies that ideas are the driving force behind development and change in science. Each individual makes his own choice, the genius of scientists is what counts. A *materialistic* viewpoint implies that the material base governs the advancement of scientific knowledge. Scientific activity reflects the special interests of those who are in control of the means of production (Widberg 1978, pp2–3). Both Harvey and Widberg are in favour of a dialectic materialistic view.

But it is also possible to argue that in geography paradigms, or rather schools of thought, have not succeeded each other as Kuhn's model suggests, but, to a large extent, continue to exist in parallel, whilst the new schools slowly absorb the older ones leaving some former contradictions to linger on within the new structure. Figure 3 suggests that some concepts survive after basic shifts in the discipline have take place.

Figure 7, like Johnston (1978, 1979), illustrates the existence of parallel schools of thought within geography since World War II. Johnston (1978) concludes that Kuhn's concept of a paradigm pays little attention to the nature of conflict within a social science. 'There is little evidence either of large-scale disciplinary consensus for any length of time about the merits of a particular approach or of any revolutions that have been entirely consummated. Certainly the quantitative and theoretical developments have had a major impact, but there are many residuals of the earlier regionalism . . . The failure to fit Kuhn's model to recent events in human geography leads to the conclusion that the model is irrelevant to this social science, and perhaps to social science in general. (Johnston 1978, pp199–201).

From this we may draw the conclusion that the dichotomy between a dialectical and a mechanical understanding of the history of our subject is an interesting academic study, but the truth as we see it lies somewhere between the contradictions. Shifts of major importance do occur, but they seldom encompass the whole scientific community — old ideas and concepts remain with us to a large extent; new discoveries may sometimes have the character of mutations — but usually look more like rephrasings of old truths.

To analyse the dichotomy between an idealistic and a materialistic view we have to discuss social research and values a little more closely and, in particular, try to get a better understanding of the debate between positivism and critical philosophy.

CHAPTER FOUR

POSITIVIST AND CRITICAL SCIENCE

'True knowledge for mankind is like sunshine for topsoil' as N F S Grundtvig, the Danish priest, writer and educator once said. The 'true knowledge' is to be produced by science through research. The prestige and support which aids research stems from the general belief that research is an almost objective, value-free activity. Most people define research as an uncompromising search for truth, an activity which is unaffected by the emotions, beliefs, attitudes and desires of either the practitioner or of the society in which the research takes place. When research workers pursue their results as if they were political questions, many people will conclude that there must be something wrong with the research and its prestige sinks. Social scientists who work with less exact or less clearly definable data become special targets for such attitudes.

Natural scientists, as a rule, work with discrete materials which lend themselves to experiment and objective analysis. Social scientists have a weaker data base, are personally involved in their problems, and can seldom experiment with their material. The question of objectivity is therefore most acute for social scientists. The results of social science research are often regarded by the community as qualified viewpoints, whereas the conclusions of natural scientists are received to a much greater extent as truths. The discussion on the role of values in the social sciences which has developed in recent years is therefore important, not only for research activity as such, but also for relationships between scientists and the general population and for the community's regard for its scientists.

Positivism

Scientific and philosophical discussion in recent years has produced two chief categories of *metatheory* (superior theories or theories about theories): positivism and critical theory. *Positivism* is connected with the naturalistic–pragmatic trend in modern thought, and *critical theory* with phenomenology and hermeneutics. At the present time positivism dominates the English-speaking world and Scandinavia, while critical theory prevails in Germany, France and the Spanish-speaking countries.

'Both trends are in a sense more "climates of opinion" than definite schools of thought. There are far more discussions within the trends than between them,' says Hans Skjervheim (1974, p213).

Positivism, in some connections called *empiricism,* has the central thesis that science can only concern itself with *empirical questions* (those with a factual content), and not with *normative questions* (questions about values and intentions). Empirical questions are questions about how things are in reality. In this context 'reality' is defined as the world which can be sensed. This means that science is concerned with *objects* in the world. The *subject*, or subjects for which there is a world, or worlds, are excluded from the field of interest.

Disregarding the question as to whether what we can sense as objects comprises the whole of reality, it could be said that empirical questions are about what a thing is, but normative questions are about what a thing should be. 'How *are* the available food resources distributed between the inhabitants of the world?' is an empirical question. The corresponding normative question would be: 'How *should* the available food resources be distributed between the inhabitants of the world?'

Positivism holds that, since we cannot investigate such things as moral norms with our senses, we should keep away from normative questions; we cannot justify our tastes scientifically. Science can describe how things are, and experimentally or by some other measurement, discover the association of causes which explain why things are as they are. The research worker can, given his knowledge of contemporary associations of causes, forecast possible developments in the future from given propositions. But science cannot from 'is' statements draw conclusions about 'should' statements. Ideally, science is value-free, neutral, impartial and objective. When the scientist gives valuations, expressing 'should' statements, he is no longer a scientist, but possibly a politician.

Another major aspect of positivism is its emphasis on the *unity of science*. Scientific status is guaranteed by a common experience of reality, a common scientific language and method ensures that observations can be repeated. Since science has a unified method, there can only be one comprehensive science. The common method is the hypothetic–deductive method and the model discipline is physics. The language which will make a unification of science possible is the physical language or *thing language*. The ultimate aim is, in the words of Rudolf Carnap, to construct 'all of science, including psychology, on the basis of physics, so that all theoretical terms are definable by those of physics and all laws derivable from those of physics' (cited by Skjervheim 1974, p222). The poles and the system of latitude and longitude are the only special definitions which must be made before pursuing geographical research. It follows from this that disciplines are to be distinguished from each other by their object of study, and not by their method (Gregory 1978, p27).

The concept of positivism was established by Auguste Comte during the 1830s in France. The concept began as a polemical weapon against the 'negative philosophy' prevalent before the French Revolution. This was a romantic and speculative tradition which was more concerned with emotional than with practical questions and which sought to change society by considering utopian alternatives to existing situations. The positivists regarded such speculation as 'negative' since it was neither constructive nor practical, it showed that philosophy was an 'immature' science. Philosophers, like other scientists, should not concern themselves with such speculative matter, but should study things they could get to grips with: material objects and given circumstances. This approach was to be recommended as 'the positive approach.' Comte himself wanted to direct the development of society, but stated that the nature of positivism is not to destroy but to organise. An organised development should replace the disorder created by the Revolution. Free speculation, or systematic doubt, as defined by René Descartes (1596–1650), was identified by Comte as the *metaphysical principle*. The word metaphysics derives from the works of Aristotle as an expression for the chapters which followed the physical (or empirical) parts of his work. Metaphysics was later redefined as that which lies outside our sense perceptions or is independent of them. Positivists regarded metaphysical questions as unscientific. Comte held the metaphysical principle (systematic doubt) responsible for the French Revolution, which had started in emotional enthusiasm, to tear down the feudal structure of society, but had ended in despotism. In a positive society scientific knowledge would replace free speculation or make it unnecessary.

Many would argue today that this combination of conservative political ideas with a rigid definition of what is, or should be, a scientific approach is typical of positivism, but it would be wrong to use Comte as the only example of positivist thinking. Politically, the most prominent followers of positivism have been socialists and liberals. There are many examples of how positivism as a scientific ideal has had a liberating effect on science and society, and it has almost always worked towards democracy.

Positivism played a progressive role in the sense that it strongly opposed the belief that the Bible was authoritative in scientific enquiry. During the Middle Ages religion had dominated what there was of intellectual life and metaphysics was its prime concern. During the Renaissance there was renewed interest among scholars in earthly matters. When in doubt why not research the matter yourself rather than look it up in the Bible or consult Aristotle? A range of taboos against empirical investigation were broken down, and it was now possible, for example, to perform an autopsy on a corpse. The data collected, especially through the 'great discoveries' of Columbus, Magellan and others, and from astronomical enquiries helped to raise science up as an independent school of thought, separate from religion.

In seventeenth-century England, acute political strife intermittently favoured independent research. Francis Bacon formulated the scientific method of induction based on factual data derived from the evidence of the senses. John Locke formulated the basic positivist principle that all knowledge is derived from the evidence of the senses: what is not derived from the evidence of the senses is not knowledge. Reliable knowledge can only come from basic observations of actual conditions. To be scientific is to be objective, truthful and neutral.

Comte, who later defined positivism as a scientific ideal in line with Locke's principles, believed that alongside the natural sciences there should also be a science of social relationships (which he called *sociology*) to be developed on the same principles. As natural sciences discovered the laws of nature, so scientific investigation of communities would discover the laws of society. He admitted that social phenomena are more complex than natural phenomena but believed strongly that the laws governing society would eventually be discovered and that subjective elements in research would be eradicated. This belief is central to Comte's proposition that social development takes place in three stages: (1) theological when men explain everything as God's will; (2) metaphysical, and (3) positive when causal connections are discovered between empirically observed phenomena.

As was shown in Chapter 3, the publication of Darwin's *Origin of Species* was a major boost for positivism as a scientific ideal. The great stress laid by positivism on empirical data and replicable research methods enabled a marked development of science during the nineteenth century. Because metaphysical questions came to be regarded as unscientific, science developed its own objectives which were apparently free of belief and value postulates. Positivism tends to be anti-authoritarian in so far as it requires us not to believe in anything until there is empirical evidence for it and it can be investigated by controlled methods. We should not therefore accept authority because it is authority, but only give credence to things for which there is scientific evidence. This sceptical attitude, and the consequential search for certainty, has naturally enough brought conflict for positivists and led them into confrontation with dictatorial regimes.

In the 1920s a group of scientists known as the '*logical positivists*' was founded in Vienna. The leading philosopher of this 'Vienna circle' was Rudolf Carnap. The group pursued a modern development of positivism, and extended its fundamental principles by arguing that formal logic and pure mathematics as well as the evidence of the senses provide sure knowledge. The logical positivists were opposed to everything that smacked of metaphysics and unverifiable phenomena. They therefore became bitter opponents of Nazism, which they saw as a mixture of irrational prejudice and ideological dogma. 'Positivist' became a term of abuse in Nazi Germany, and was applied to Alfred Hettner, among others (van Valkenburg 1952, p110). The Nazis wanted research to be based on their own ideological attitudes. The leader of the Vienna circle, Moritz Schlick, was murdered by the Nazis and its other members driven abroad. As long as science uncompromisingly seeks to discover the truth, it will endanger regimes based on systematic lies and ideological postulates.

Criticisms of Positivism

It has been asserted that positivism is not so anti-authoritarian as it claims to be because it seeks authority from the methods of natural science. This may lead positivists into thinking that there are technical solutions to all problems — an essentially conservative standpoint. The wish to be free from value-judgments may lead scientists to build ivory towers for themselves wherein research is excluded from any discussion of objectives and decisions. Positivist research workers might restrict themselves to describing how things are and how they will develop if they continue on the same track as now. Critics maintain that the prestige of modern science creates an aura of inevitability around these 'are' statements and tendencies.

Others go further in arguing that value-free research is impossible. Subjective elements will intrude in many stages of the research process especially at the stage when research workers choose a topic for study from the many available. We can, for example, guess that a research worker, starting from his own well established and strong opinions as to what the distribution of the world's food supply *should* be, will choose to investigate the empirical question as to how the food supply is *actually* distributed. Even if the research worker does not deliberately consider what the distribution should be, it would be difficult for him wholly to exclude his own views at the stages of problem formation and the interpretation of the results. It is also obvious that once results are available, the description of the existing distribution will influence the views of many decision makers, as to what the distribution should be. In this process we can say that scientific activity is itself shaping reality. It is no longer a passive observer.

Other critics maintain that the belief of positivists in the unity of science is wholly unrealistic; positivism allows its view on the logic of science to influence its conception of the content of science. We should not oversimplify things by laying down rules as to how science should function without taking account of what actually happens within the livelier research traditions. The fact is that several main lines of scientific enquiry, including important schools in psychology and social science, do not show any sign of developing towards the ideal of a unified science. Ideas about unified systems of concepts belong on the drawing boards of the theoreticians of science; the inner drift in science itself is to give each area of investigation its own form of expression. It is even possible to argue that each and every scientific paradigm or school of thought is a form of cognition, an agreed approach to the analysis of the world. Teaching a discipline consists in teaching its current forms of cognition. When, for example, one learns to see things geographically it is not reality itself one learns, but a perspective on reality. What the positivists are trying to do is to base all knowledge on methods which are carried over from the natural sciences into the human sciences. They have developed a specifically technical conception of science, and try to exclude scientific traditions that do not adapt to their recommended technical language and methods. A typical example is Schaefer's rebuttal of the chorological viewpoint and his efforts to make geography into a *spatial science* (see pp59, 140–3).

Wilhelm Dilthey considered that while we *explain* nature, we *understand* (*verstehen*) social life and human intentions. The distinctive properties of social science which have developed out of the Diltheyan tradition are based upon his conceptions of meaning and *verstehen* which necessitate a twofold division of the sciences into natural sciences and cultural sciences (Dilthey's term *Geisteswissenschaften* includes both mental and spiritual conceptualisation). The Diltheyan tradition supports the use of empirical methods within the natural sciences but does not agree with Comte and Carnap that the social/cultural sciences should copy the methods of natural science. When social scientists borrow models of system building from natural sciences they must treat all their elements as objects. When the mind is treated in this way it is materialised, either directly by being conceived of as the thing that thinks, or indirectly by being considered as a relation between objects in the world. The same applies to behaviour. 'What is lost to view is mind as correlate of the world of objects, as that for which there is a word constituted' (Skjervheim 1974, p216)

At the extreme, we might talk about brain processes instead of ideas or images, but we cannot study human behaviour in the same way as we study animal behaviour. The difference is that men have *intentions*. Intentional expressions such as 'to imagine something,' 'to say something,' 'to believe something,' 'to love somebody' cannot be translated into the 'thing' language of the natural sciences: they cannot be understood as objects as seen from the outside. If you study your fellow men as physical objects you will not get to grips with their intentions. It is better to build bridges to other people by breaking down the barriers between the observer and the observed and creating an intersubjective understanding. This is the principle of *subjectivity* in social science, which says that behaviour has to be studied and described in terms of the actor's orientation towards the situation.

This position is connected with Immanuel Kant's criticism of empiricism as a philosophy. As a man of deep religous belief, Kant worried about what he saw as 'nihilistic implications' of empiricism. Neither empiricism nor positivism leave room for a God, nothing is *a priori* certain. Kant's need to ordain certain main features of truth was satisfied by his doctrine of 'categories', in which he claimed that if there is no prime cause for the content of reality, the form of reality must be primarily given. What the well of consciousness is filled with is an empirical question, but regardless of what is put into it, the content is shaped by the form of the well. Mankind cannot know how things are in themselves, we only recognise things as they take shape through our senses and by conceptualising them. The reason why pure chaos does not reign within the field of scientific activity is that human perception and reflection are built upon common categories within which impulses are classified (space, time, cause). Kant classified the branches of science, as we saw above, into three corresponding categories (p15).

Georg Wilhelm Friedrich Hegel made this even more difficult when arguing that the categories we use for classification and thought (structure of concepts, speech) are not fixed for all time, but are historically and socially conditioned. Often we cannot gain a full understanding of a thing by merely studying it on its own. We must also consider its antithesis — its opposite. A geographer, for instance, is often in a far better position to describe his home area however well he knows it, after he has had extensive experience of foreign parts. We might cite the comparative method of Ritter and Humboldt. Insight might be improved through a constant consideration of the antithesis which throws new light on the thesis. The development of knowledge on the Hegelian model is usually described as a *dialectic*. As we increase our insight into the thesis by contrasting it with its antithesis so we heighten our understanding of the synthesis which, in its turn, enables us to set up new theses and antitheses against each other, etc. This process does not achieve a permanent form of knowledge, correct for all time. Knowledge and science are not like a steadily growing anthill of pine needles, but are better defined as continually changing and deepening processes (Kuhn 1962, 1970).

Gunnar Olsson (1975, p29) suggests that the crucial point about dialectics is that understanding and creation themselves are moved by dialectical transitions, truths are relative, statements can be designated as 'true' only at a given point in time and, in any case, can be contradicted by other 'true' statements. 'While conventional reasoning knows only the either-or distinction of the excluded middle, reality knows the both-this-and-that relation of dialectics and many-valued logics. Since contradiction is not external to reality but built into its structure, the language in which reality is discussed should itself have the same characteristics of internal negation. It is in this sense that reality is dialectical, for both reality and dialectics are governed by perpetual processes of internal tensions and not by stultifying juxtapositions of opposites. As a consequence, both the empirical and the logical bases of our theories should be open to conceptual change. But the aim is not to falsify for the sake of rejection. It is rather to falsify for the sake of creative understanding. This is possible, because dialectic movement is not inference but deepening of concepts' (Olsson 1975, p28–9).

Marxism

Karl Marx used and developed the dialectical theory further, and for this reason he has often been identified as one of the founders of critical philosophy. David Harvey, however, in *Explanation in Geography* (1969, p47) classified Marx as a positivist, in line with Auguste Comte, Herbert Spencer and John Stuart Mill. Later Harvey changed his view of Marxism, suggesting that there is an essential difference between positivism and Marxism, although there are certain things which they have in common (1973, p129). Other observers consider that the Marxist dialectic stands on its own without strong associations with other philosophies.

Marxism and positivism have in common an analytic method and a materialist base. For Marx the intellect was more the product of material conditions than their foundation. The material world and men's behaviour therein was the base; thoughts and ideologies formed the superstructure.

There have been lively discussions among Marxists as to how we are to understand the view that the ideological superstructure reflects the materialist base. This can be understood in the sense that the main directions in history are determined by material, concrete conditions. When these basic conditions are identified by Marxist analysis, we have a sure way of orienting ourselves in relation to the future, so certain in fact as to provide a basis for positive political action. It is not the ideas that change the world, but the development of actual reality which changes the ideas. In its crudest form this idea developed in *vulgar Marxism* which is based on economic determinism, implying that feudal society *had to be* followed by capitalist society which in its turn *will be* replaced by the dictatorship of the proletariat which will finally develop into a communist society.

While vulgar Marxism has had a profound influence on the political scene, Marxist scientists in general will cite Marx and Engels in emphasising that the relation between the base and the superstructure cannot be understood so simply — it represents a dialectical situation, a reciprocal influence between the material and the ideological. Friedrich Engels stated that it is through interaction between all these factors (political, legal, philosophical, etc) that economic development takes place as the necessary outcome of a number of accidental occurrences.

According to this view, individuals and their ideas can have *some* influence on the development of society.

Whereas there might be some doubt as to the deterministic character of the grand Marxist theory that society develops in stages in accordance with the developments in the factors of production, it is quite clear that Marx refused to accept that the scientific laws of society were eternal. This view contrasts sharply with the claim of positivist science that scientific laws are universal in space and time. Engels pointed out that 'to us, so-called economic laws are not eternal laws of nature but historical laws which appear and disappear' (cited by Gregory 1978, p73). Marx was particularly critical of the economists Adam Smith and David Ricardo; he considered their outlook on reality to be unhistoric and non-dialectic. Society has inbuilt conflicts which will resolve themselves by change both in practice and in theory.

A positivist aspect of Marxism is the theory of realistic knowledge. Knowledge means cognisance of an objective truth: it regards the world as concrete. It consequently rejects Kant's philosophy. But at the same time it is clear from the Marxist point of view that science and scientific development must be understood in relation to social reality. The materialist base influences science which will have difficulty in maintaining a real objectivity. For this reason, the relative autonomy of science is important because its guarantees a higher degree of objectivity. The consequence of this seems to be that a theory cannot necessarily be recognised as correct because it supports a standpoint of the proletariat, or false because it is in conflict with the interests of the workers. Science and scientific values must be regarded as autonomous. This stance, however, does not wholly support the positivist ideal of a value-free and neutral science. Harvey (1973, pp129–30) states that the essential difference between positivism and Marxism is 'that positivism simply seeks to understand the world whereas Marxism seeks to change it.' The political aims of Marxist science are also expressed by Richard Peet (1977, p21): 'Marxist science begins with a material analysis of society, proceeds through a critique of capitalist control of the material base of society, and proposes solutions in terms of social ownership of that material (economic) base . . . the political objectives of Marxism provide a common scientific purpose. As a holistic, revolutionary science, Marxism provides a firm theoretical base for the radical movement in geography.'

Marxism also proclaims the unity of science, but the basis for this unity is not so clear-cut as in positivism. Engels leaned to the natural sciences when he maintained that the gradually increasing similarity between theories of social science and those of natural science would lead to the accommodation of nature and society within a unified philosophical perspective; we may here trace the ideas of social Darwinism. In his youth, Marx also wrote about a unified science which comprised nature, society and human psychology, but while positivists maintained that a unified science should be based upon the methods of the natural sciences, Marx considered the philosophy of the social sciences to be potentially far superior to that of the natural sciences. Therefore the eventual fusion of the two fields of study would come about through the socialisation of the natural sciences (Harvey 1973, p128).

Critical theory has been widely associated with Marxism because it was adopted by the anti-authoritarian and mainly socialist youth movements of the late 1960s. But critical theory is not a political theory: it does not answer political questions. The only political implication is that critical theory is even more anti-authoritarian than positivism in that it subverts our acceptance of the view that the methods of physics will lead us to the solution of all scientific problems. The 'systematic doubt' which is brought into operation again has very little in common with Marxism — at least in the way that Marxism is practised. The young Marxist generation soon returned to orthodoxy, preferring to discuss how Marx should be interpreted correctly rather than to develop new social ideas based on free speculation and systematic doubt.

Subjectivity and Objectivity

Critics of positivism maintain that metaphysical assumptions cannot be excluded from science. Facts are not facts in themselves: they represent those parts of reality which can be appreciated with the concept apparatus available. Facts are facts only in relation to a given scientific aim which is itself structured by the values intrinsic in society. The scientific process is therefore restricted by the environment in which it takes place and is constrained within the limits of the perception of research workers. This view of science implies that theories are preconceived and therefore determine observations rather than providing explanatory structures after empirical observations have been made. Consequently, the proponents of different theories see the world in different ways and the comparison of different theories by submitting them to objective tests, which any qualified scientist could carry out, is totally impossible. We may have some difficulty in explaining the progress of science in this rather frustrating situation. Some offer a dialectical solution; scientific discourse is regarded as a structure in which theories and observations progressively transform one another (Gregory 1978, pp57–8).

Although some regard these assumptions as valid throughout science, most advocates of critical theory would stress the difference between *natural* and *social sciences*. In a practical way, Immanuel Kant himself carried out empirical research in physical geography and did not regard this as being in conflict with his rejection of empiricism as a general philosophy for all sciences. Natural scientists have not regarded the discussion of positivism as an urgent task, for the research that they do is independent of their philosophical stance. They have been able to construct a free-standing scientific language — the 'thing' language — which satisfies their need for precision. This language has admittedly been established through the subjective appreciation of nature by scientists, but this creates few problems as long as there is a measure of agreement on the experience of the senses within the scientific world. Nature can only answer in the language in which the question is put — which is the language of science. Therefore mistranslations can only occur through the faulty use of accepted methods and language.

Many scientists would say that critical philosophy has little relevance for theory in natural sciences, but that it is more relevant to the practice and application of science. It may be possible to pursue more or less objective research within natural sciences, but the results of science are applied by society in such a way that science becomes a tremendous force which transforms the community (Ackerman 1963, p430). Value judgements are implicit in the application of scientific results, and consequently the natural scientist cannot avoid making up his mind about them. Even the priorities given to different types of research are connected with values, it is not irrelevant whether we give priority to nuclear physics or to ecology.

In social science *subjectivity* or the problem of values is deeply involved in both theory and practice. Concepts in social science are related to human evaluation; a typical example is the concept of *natural resources*. A natural scientist might study coal as a thing in nature, but in social relationships coal is a resource which must be translated as 'something of value to man.' Coal is only interesting to social scientists because it is a resource and forms the basis for coal-mining districts. But coal has not always been a resource, and not all coals found in nature are resources. The market situation and the cost of extraction may lead to closure of a coal mine even if there is a lot of coal remaining. The human evaluation which creeps in makes the use of the 'thing' language difficult in social science.

Another problem is that ordinary language must be used widely in the social sciences, because it is spoken by the objectives of the research and also because there is a close affinity between daily speech and technical language. When, for example, we ask a commuter for his opinions about his own pattern of life we must take account of the concepts of the interviewee. A problem arises when we define our concepts 'scientifically' before embarking on a research project and later use these definitions to interpret our results without critically evaluating what happened during the investigation. The person who has been interviewed may have placed a wholly different construction on the concepts. The dialectic between subject and object, which forms a major aspect of the philosophy of Hegel, actually seems to take place in the social sciences. It is impossible for the research worker who is investigating social phenomena to regard them as objects which are wholly external to himself. The subject is himself part of the object: the social scientist acts as part of the society he studies.

This appreciation lies at the root of the *hermeneutic* (Greek *hermeneuin* = to interpret) tradition within social science. This tradition tries to reveal expressions of the inner life of man by *verstehen* (a German word which means something between understanding and empathy), by putting oneself in another man's shoes. Max Weber distinguished two kinds of *verstehen:* observational understanding and motivational understanding. We have direct observational understanding of the meaning of spoken and written propositions, facial expressions, emotional expressions, etc — these things can be studied as objects, just as things can be studied in natural sciences. Motivational understanding, on the other hand, takes place when we impute a motive to someone in order to understand and explain why he is doing what he is doing. We cannot, however, impute a motive until we have understood what he is doing, that is, motivational understanding is based upon observational understanding. A scientist cannot achieve a motivational understanding merely by observing the people he is studying as if they were objects. Instead he must try to place himself on an equal footing with the people he studies; he must try *to be* those studied so as to be able to see things from their point of view. This method of subjective identification contrasts strongly with the methods accepted by the positivists. The dialectic between the subject and the object leads on to what has been termed *double her-*

meneutics. The social scientist explains whatever he has understood about the beliefs and attitudes of his fellow citizens in his writings, but the people he writes for are more or less identical with the people he is writing about. If the reader understands what the writer has understood, he may be led to change his attitudes or actions. The new attitudes can then be investigated again by the social scientist, explained and understood *ad infinitum*. The meaning ascribed to the one constantly mediates the meaning ascribed to the other through double hermeneutics or the hermeneutic circle (Gregory 1978, p61; Skjervheim 1974, pp298–9)

Since our knowledge of the world is established through the concepts of science, reality and knowledge can be changed through human reflections and practice. Reality, as it is constituted by scientists, can therefore be transformed into something else. Critical theory implies that science should take an interest in these possibilities of change.

Positivists maintain that critical theory has a metaphysical origin which is irrelevant in a scientific context. Kant's question as to whether things in reality are different from the picture we have of them through sense perception may be of philosophical interest. It is not the duty of science, however, to reproduce reality, but only to describe and explain how things appear through the evidence of our senses. In practice, there is a great deal of similarity between the sense perceptions of different people. Differences depend to a great extent on imprecise or poorly developed concepts. Positivists further argue that the adherents of critical theory confuse scientific and interesting human problems. Science should only study what can be treated as objects in the world, if human intentions are not susceptible to such scientific enquiry, scientists should refrain from analysing them.

Most positivists will agree that value judgments play a major role in the research process and that science ultimately takes part in the shaping of society. Positivism is, however, not a description of how research is actually carried out; it represents an ideal of how it should be done.

Practical Consequences for Research

The chief effect of the conflict between positivism and critical theory on practical social research has been to enliven the discussion about the influence of value judgments on research activity. The thoroughgoing positivist standpoint stresses the importance of reducing the value element as far as possible. Research workers would be wise to refrain from problems on which they hold strong opinions which might influence the research process or cause faults which would reduce the value of the findings. Ultimately, we should totally avoid subjects where strongly divergent views are held so as to escape being implicated in political conflicts.

Others consider that we should not avoid contentious questions or even questions in which we have strong interest ourselves; in these circumstances we should be on the alert for outbreaks of subjectivity. This can be achieved by maintaining careful accuracy in description and by carefully describing our methods so that other research workers can confirm our results.

Some scientists would not regard this as sufficient. As well as taking care to avoid subjectivity, research workers should express their personal viewpoints clearly so that their significance can be evaluated and accounted for in the overall consideration of the results of the research.

A contrary view is that research workers should take up exactly those questions in which they feel deeply involved. The influence of value judgments can be positive when it motivates us to greater efforts. Research workers should express rather than try to eliminate the influence of values on their work so that these influences, being observed, will enable different views within the research environment to correct each other.

Some would go as far as to say that research workers should use science to fight for their values. If the 'value' is a political viewpoint of the research worker, he may set the problem and publish the results of research in order to support his political viewpoint. This attitude is supported by the argument that research, according to the ideals of the positivist, does not consider what the community could have been, only what it is. Such research supports the existence of, and consequently provides unconscious evidence for the values of the surrounding society. The research worker who does not share these values must therefore be free to pursue research with the intention of changing basic values and therefore the nature of society (cf. Harvey and Peet's views, pp72 and 84).

There will undoubtedly be different views as to how we should tackle the practical research problems created by value elements. We may agree that neither research ethics nor objectivism demand freedom from values; the influence of values must be accepted as an inescapable element in research. Research ethics therefore become closely associated with the degree to which value judgments are clarified: 'We need viewpoints and they presume valuations. A "disinterested" social science is from this viewpoint pure nonsense. It never existed and never will exist', says Gunnar Myrdal (1953).

The practical consequence of such an appreciation is that the research worker must be free to choose research projects which he or she regards as critical. This implies that scientists themselves, by and large, should have the power to decide how the funds for research should be distributed. Certainly each community will have a research policy but this should not be carried so far that politicians and those assigning tasks at the administrative level outside the scientific environment, actually determine research tasks in detail. This approach to the freedom of research is widely shared among scientists, although in many countries the reality does not confirm to this ideal. Positivists have always held this view, and even Marxist theory stresses the importance of the relative autonomy of science, although in this respect the practice of communist countries often deviates from theory. If the description of scientific activity contained in critical theory should become widely accepted, however, it might be difficult for science to retain a degree of autonomy even in democracies, for it is much easier to justify the demand for free enquiry when it is held to be value-free, than when it is admitted to be guided by outside values.

Politicians and the people in general are just as qualified to make value judgments as are scientists. If science is admitted to be a force which transforms society, and the ivory tower of science becomes a governing structure of society, it must clearly be brought under democratic control. It is impossible for science to maintain both priviledged autonomy and a commanding position. Up to now scientists have generally regarded their autonomy aṣ more important than their power, and have consequently tried to separate their political from their scientific activities. As both a scientist and a politician, the author appreciates that he has been in a privileged position in his political activity, since scientists are generally better trained to formulate problems and to sift out the essential elements from volumnious official documents than is the ordinary citizen. For this reason the influence of science and scientists may be much more of a problem for democracy than the influence of society is for science.

Value elements are most significant in the first phases of the research process when research problems are chosen, set and defined, and also in the concluding phases of research when the results are interpreted and presented. In the intervening data-handling phase, value elements do not play such a large role; they primarily concern the classification of data.

When interpreting and presenting research the problem arises as to how far to carry the conclusions and where to cut short. Is the assignment completed when the relationships shown by the data have been described, or should we also use our imagination and theoretical insight to add something about the circumstances under which these relationships will change or disappear? This problem is particularly acute for social scientists, where the possibility of changing direction depends on human decisions.

In fact, everyone agrees that research workers should give a clear explanation of the conditions for and the circumstances leading up to their results; this is essential to scientific integrity. The majority will also agree that, within the social sciences, the relevant environmental conditions may include the existing political and social system. It would also be reasonable to consider those aspects of the social system under investigation which, if they were changed, could change the situation which is under investigation. There is, however, disagreement as to how far one should question the normally accepted environmental conditions. The majority view is that scientists overreach themselves if they try to demonstrate the correctness of a particular political viewpoint and thus erect a new kind of teleological philosophy of science (cf. Ritter's teleology). The boundary between science and politics is, however, often difficult to trace. Scientists who feel strongly that their society is governed by an élite group may feel justified in acting as 'counter-experts' on behalf of the less articulate ordinary people. They must then decide whether they are justified in using their status and employment as professional scientists on such an assignment or whether to do it in their spare time as well informed ordinary citizens.

It is also generally agreed that it would be an untenable situation in universities if a piece of research work, or an examination question, could only be assessed by examiners who shared the political outlook of the candidate. We all recognise, however, the difficulties in giving a full and objective judgment of a piece of work in the social sciences.

Idiographic Traditions in Geography

The question of the role of values in research has only been debated by geographers relatively recently. One of the reasons for the lack of discussion has been that geography has had a stronger association with the natural sciences and is more 'concrete' than the other social sciences.

The traditional view is that the natural sciences are more objective than the social sciences, but there are, however, considerable differences between an 'abstract' and quantitative science like nuclear physics and a 'concrete' and less quantitative science like geology. The smallest objects in physics, such as the electron, are not capable of observation, so we may say that they are 'abstract' because they only exist in theory — a theory nevertheless based on numerous and exact quantitative measurements. The problem of objectivity at this level may be more acute in physics than in geology, but any research worker who is studying the influence of one variable factor on a single research object, who can measure this influence exactly and present his results numerically, will have very few problems concerning objectivity.

The fact that geography in its intermediate position within the sciences includes elements of natural science is *one* reason for the lack of a positivist debate in the subject. Another reason is the idiographic tradition in geography which has been supported by geography's strong links with other idiographic disciplines like geology, biology and history. Physics — widely regarded as the model science — is characterised by the hypothetic— deductive method, well developed quantitative techniques, and the establishment of laws. Physics is the model for scientific activity in chemistry, geophysics, medicine and, to some extent, in the physiological sections of zoology and botany. The systematic branches of botany and zoology have traditionally been confined to a classification and description of species. Only during the last three decades has an increased interest in ecological relationships led to discussion about the formulation of laws and methodological issues in these branches of the biological sciences. This development runs parallel to the 'quantitative revolution' in geography. We may note that geography has its strongest contacts with the systematic parts of biology through plant and animal geography.

Geology is the natural science which has traditionally been closest to geography. The research methods of geology are described *inter alia* by G G Simpson (1963, p46) as different in nature from those of physics. He associates geology with what he calls historical sciences. Geology refrains from the formulation, verification or rejection of hypotheses through experiments and the establishment of universal scientific laws. It describes and clarifies 'concrete,' simple phenomena and puts them into a geological chronology and classification.

Similar comments may be made concerning traditional views on history as a discipline. W H Dray, for instance, argues that 'History . . . seeks to describe and explain what actually happened in all its concrete detail . . . Since . . . historical events are unique, it is not possible for the historian to explain his subject matter by means of covering laws' (Dray 1957, p45). Whilst, in physics, individual phenomena and their combinations cannot be presented as really new objects, historical phenomena, singly or in combination, can hardly ever be subjected to a uniform form of measurement; they exist as truly new phenomena. The laws of the natural sciences may be regarded as having unrestricted universal validity but historical generalisations are not valid for all times and places.

The views expressed above would be supported by the majority of historians although there are some who disagree. O F Anderle, for instance, expresses a positivist view: 'Though the canonisation of the idiographic method is supposed to establish the independence of the historical from the natural sciences, it really just commits the former to an earlier stage of the latter. Descriptive historiography is not a new science with an independent method but just an antiquated form of natural science' (Anderle 1960, pp40–1). Oswald Spengler and Arnold Toynbee searched for generalisations or historical laws as part of their attempt to write 'large-scale' history. The ideas of Marx have been fruitful in economic history for, although only a few research workers have followed them slavishly, they have had a major impact as law-formulating elements in that subject.

The majority view on the significance of individual occurrences which has dominated these neighbouring sciences has, as we have seen above, also been important in geography. The determinism of the nineteenth century proved to be rather an unfortunate attempt to establish laws and to use hypothetic–deductive methods. Its lack of success led to scepticism towards nomothetic approaches in geography. Geographers therefore turned their backs on that major element of positivism — the unity of science with one methodology whose results are not modified by time and space.

Even during its most idiographic phases geography has been prepared to accept certain 'laws' or generalisations. These generalisations have not been regarded as truths in themselves but rather as tools with which to measure truth. We may for instance suggest a law which gives an ideal model for the distribution of towns in a region. This 'law' may then be used as a tool to evaluate the many discrepancies and to compare the distribution of towns in one region to that in another (cf. Chisholm's arguments on geographic laws, pp66–7). This seems to be identical with Max Weber's (1949, p80) views on social science methods and law formulations. Weber was one of the founders of the hermeneutic tradition. He is regarded as a forerunner of the school of critical theory represented by Karl Mannheim,

Herbert Marcuse and Jürgen Habermas, all of whom emphasised the need for distinctive methods and theories in social science. There is accordingly a considerable degree of consensus between the traditional geographical viewpoint and critical theory with regard to the universality of scientific methods and laws. For this reason thoroughgoing positivists have not been numerous in departments of geography, and the debate between positivism and critical theory had only a limited interest for the subject.

After the 'quantitative revolution' this situation changed. Schaefer (1953), for instance, attacked 'the old parallelism between history and geography,' refused to accept the hermeneutic method of *verstehen* (empathy/understanding) as scientific, and insisted that 'science begins only when the historian is no longer a historian in the narrow sense and tries to fit his facts into a pattern.' He wanted geography to establish itself as a law-seeking discipline and use the 'scientific' method. Since Schaefer had had his training in the Vienna circle of logical positivists (as pointed out by Gregory 1978, p32), his arguments are more clear-cut than those expressed by most advocates of the quantitative school. As a consequence of the 'quantitative revolution', however, many geographers came to consider that science really provides a unity, that geography ought to become a law-seeking science and should use the hypothetic–deductive method. In this way they accepted a major element of positivism.

Geography as an Empirical Science

There are other elements in positivism which have been found in geography since Darwin's time. Geography has clearly been defined as an empirical science. Its data are concrete, it studies what really exists (especially what exists at the present time). The question as to what reality could or should consist of is regarded as unscientific, being instead political or metaphysical in nature. This view may well present fewer problems for geography than for sociology and other social sciences. That part of the data base of geography which concerns natural conditions is concrete and can be unambiguously defined by natural science terminology and methods of measurement. As long as we do not get involved in Kant's philosophy, definitions are indisputable, provided that the terminology and methods are used correctly.

Much data in human geography are similar to data in physical geography in that objects which appear on a map or air photograph including patterns of settlement, lines of communication and elements of land use are as concrete as the data used by agronomists, geologists and biologists.

Although some other material in human geography is not so clearly manifest in the cultural landscape, most of it is concrete and *measurable*. We may consider, for example, the numbers settled or employed at a place, transport as measured in quantities of goods carried, or numbers of vehicles or persons, the use of raw materials and energy in production and the amount of material available. These are data which are in principle as easy to handle as those which the 'concrete' natural scientists study. There is a difference, however, in so far as geographers are often interested in the general view; they look at things in a reducing glass (equivalent to the scale of the map), while natural scientists often use a magnifying glass. Consequently, it is normally rather difficult for geographers to measure their data directly with scientific precision and instruments.

We must often be content with data which we have collected through interviews, which may be affected by the perceptions of both interviewers and interviewees. We may use data drawn from statistical publications which may have been collated from a number of sources. In both cases subjective elements can introduce flaws. The definitions of science may be different from those of daily speech. Another problem may arise when officials who prepare the statistical material for publication do not understand or do not use definitions in a proper manner. A third problem is that the definitions change from one census to another; for example, the concept 'household' had quite a different meaning in 1970 than it had in 1900. A fourth problem which is also important for students who follow a development through time relates to changes in the geographical units used. Data on English counties in 1975 are not comparable with data from the 1971 census because of the major boundary changes in 1974. 'Urban area' is a reasonably precise, scientific concept, but its very definition determines that the size of the urban area normally increases from one census to another as settlements which formerly lay beyond its fringes are swallowed in the expanding town. The population growth in urban areas is due not only to the growing concentration of settlements but also partially to the increased outward movement of the towns' inhabitants. The examples show that a satisfactory definition and understanding of the data is very important. They show also that geographical data lie in a transition zone between the precise data which can be obtained through measurement in a subject like physics, and the imprecise data which social sciences and history need to use.

The critical understanding of sources is central to historical method.

This is to be expected in a subject where the data, both what are described as concrete events and the reasons put forward to explain the circumstances, chiefly consist of written records from the past. The *source-criticism* method was developed by Leopold von Ranke, who observed how every account was undoubtedly influenced by its author, and that we should be aware of this when using the source. It is therefore important to investigate the background sources for the account and the extent to which they may have been affected by the concepts of the time. Ranke also had a sharp eye for the individual characteristics of his sources and their political, social and religious points of view (Clausen 1968, p26).

Recognition of the subjective element is a basic feature of critical theory, but the growing importance of the source-criticism method is due to the demand by positivists for objective research. The development of our critical faculties enables us to reduce the subjective elements in our data, and we can reduce our subjective role in the research process by understanding it. To achieve a fullstanding objective method of historical research is neither possible nor desirable in the view of Robin Collingwood (Clausen 1968, p35). Collingwood considered that history cannot be regarded as a process separate from the research worker: historical circumstances are not natural processes. The things historians study are not only incidents which can be observed from without, but are experiences which must be relived through the research worker's own consciousness. This is wholly in line with the 'hermeneutic' method of social science developed by Max Weber and Wilhelm Dilthey by which to understand a social development is to relive it in the situation in which the original participants found themselves and see it through their eyes. The research worker should understand his role as thoroughly as an actor who is preparing for an outstanding performance. In this method, research becomes an art. Nevertheless, the subjective insight is attempting to reach an objective appreciation — by seeing the events through the actors' eyes the researcher effaces his own subjectivism. Source criticism and *verstehen* have been used by geographers, notably by the French school of regional geographers, and have reappeared in modern humanistic geography (see pp70–1).

The stress laid until recently on geography as a nomothetic, contemporary science may explain why there is so little systematic teaching of the critical analysis of sources and the hermeneutic method in geographical courses. Contemporary sources are considered to have so few problems attached to them that we need hardly study them seriously. The hermeneutic method is chiefly important when the research worker buries himself in periods or environments where the cultural background is different from his own. Western geographical research has concentrated on contemporary and local issues to a very marked extent in recent decades, and geographers have tended to ignore problems where the cultural environment is significantly different from their own. Social anthropology on the other hand, in order to develop its hermeneutic method, has sought out unknown environments as fields for its research.

Geography is similar to social anthropology and sociology in its concentration on contemporary data. Geographers, however, usually work with concrete or quantitative data, whilst sociologists and social anthropologists use 'interpreted' data — data about men's values — to a great extent. Sociologists are often concerned with the interviewee's subjective understanding of a value-loaded question. They must make certain that the interviewee understands the question exactly so that there are no misconceptions. Geographers have this problem too, but many geographical questions are easier to define precisely. Sociologists must decide whether the answer of the interviewee really expresses a subjective meaning. This is a problem which is also recognised by geographers who have worked with 'soft' interview data. We often interview someone who has not thought about our question before, and either gives the answer which he believes will be most acceptable to us or the answer which is in accordance with what he believes to be the general understanding of the question, even if this does not coincide with his own point of view. A young lad who was asked why he had moved from mid-Wales to Birmingham answered the question the easy way by saying that there were no jobs in mid-Wales, which was an acceptable answer. He concealed his personal reasons for moving: 'Things are more exciting in Birmingham and there are more girls around.' Since the data on which sociologist's work are 'soft' and the problems they decide to study are often value-loaded, it might have been expected that the traditional views of scientific theory and methodology which have dominated history and geography would have been held in sociology also. In fact positivism has had considerable significance in sociology. Important schools of sociologists and psychologists have sought laws and have used the hypothetic–deductive method to a much greater extent than geographers and historians. The discussion of positivism has consequently been much more thorough in sociology.

The Quantitative Revolution and Positivism

When geography began to draw on economics in order to articulate more formal location theories its own somewhat fuzzy empiricism was considerably strengthened and sharpened, observes Derek Gregory (1978, p40). The quantitative revolution involved the acceptance of those elements in positivism which had previously been disregarded, namely the concept that there is *one* science and *one* methodology which extends from the natural into the human sciences. By rejecting the notion that geographical phenomena are unique, the quantitative school discarded the idiographic traditions of the discipline and set out to discover universals, to build models and establish theoretical structures into which geographical reality might be fitted. The more distinguished proponents of the quantitative revolution however, make the essential reservation that their models and laws cannot be understood as if they were laws in natural science. They are measuring rods, tools which are used to test departures from geographical reality. Critics like Gregory (1978, p40), however, argue that even if it has not been possible to show that geographical phenomena are subject to universal laws, there has still been some value in regarding them *as if* they were. Some of the quantitative models and laws, he maintains, have been used as devices whose utility is measured by the success of their predictions and not by their implicit validity or truth. This approach to laws, which may be termed *instrumentalism*, was borrowed by geographers from neoclassical economics in which (according to Gregory 1978, p41) it has played an important supporting role. Instrumentalism refers to the extent to which models and laws are seen as instruments of manipulation rather than explanatory devices. Chisholm (1975, p125) refers to the same usage as *normative theory* (see p67 above). There is no common ground of agreement at the moment amongst geographers as to the degree of universality of laws and models. The author is of the opinion that few even amongst the most quantitatively inclined geographers are ready to argue that the laws and models of human geography and of the social sciences in general are unchanging and universal. Societies change, and so do the laws of society. We can divide the world into different universes: capitalist, socialist, etc, with laws that differ.

There is no common agreement either on the role of science in advocating change. The traditional viewpoint has been positivist, in line with the views of Rudolf Carnap, one of the central figures of the Vienna circle of logical empiricists, who said in his autobiography: 'All of us in the Circle were strongly interested in social and political progress. Most of us, myself included, were socialists. But we liked to keep our philosophical work separated from our political aims. In our view, logic, including applied logic and the theory of knowledge, the analysis of language and the methodology of science, are, like science itself, neutral with respect to practical aims, whether they are moral aims for the individual or political aims for a society' (Schilpp 1963, p23).

This view of science is widely held amongst geographers involved in research partly because geographical data are generally 'concrete.' In recent years the discipline has become more and more involved in applied research, especially in the field of planning (see Chapter 3 above, p67). The quantitative methods and models have, to a large extent, been developed because they are thought to have considerable predictive value. This raises the immediate problem as to whether in forecasting we should rely on the projection of current trends or try to envisage alternative scenarios. The political outlook of the research worker is very likely to affect his answer to that question. The author thinks that the important requirement of objectivity demands that we should analyse and state clearly those assumptions (or pseudo-laws) which can change or may be changed, and also the consequences which such changes may have on social development. It would be dishonest for research workers interested in the prediction of future changes not to do this.

Not all scientists will be able to attain such an ideal of objectivity. Those who support, even subconsciously, the maintenance of the existing social structure soon come to believe that the overturning of certain assumptions or pseudo-laws is unrealistic or wrong. Those who favour the overthrow of the social order, on the other hand, will easily overestimate the possibility of converting the collection of empirical laws into pseudo-laws. This position appears to be held by Gregory when he states (1978, p77) that 'the function of social science is to problematize what we conventionally regard as self-evident.'

Jürgen Habermas asserts that the relationship of the social sciences to the community is similar to that of the psychoanalyst to his patient. The psychoanalyst treats his patient by using his own intimate *verstehen* or understanding of the patient and by making the patient understand the underlying causes of his problems and that these causes can be altered. The task is to convert what the patient believes to be unchangeable constraints into pseudo-constraints. The patient who becomes conscious of these constraining conditions may be restored to health if he really wants to be healthy. The social scientists, who claim to be the professionally established reflection of society, must also reveal pseudo-constraints so that the potential opportunities for change in society are clarified. Society, like the patient, must make choices when the possibilities become clear. It may be that society, like some patients, does not really want to get better.

The parallel between psychoanalysis and social science could be interpreted so that we accept that society is sick *a priori*. By doing this the social scientist starts from a conscious value judgment that it is his duty to cure a sick society. Some social scientists would extend this analysis further, considering research should contribute to a fight for definite political values or ideologies.

Latterly, as a result of the quantitative 'revolution,' many geographers have come to regard this viewpoint as necessary in geography also. The greater emphasis on the unity of science, on hypothetic–deductive methods, on the use of 'hard' data and also our involvement in planning, made it clear to most of us that geography as a science is involved in shaping reality by explaining how situations are and what they could be like. Some geographers, however, directly attacked both the quantitative school and the positivist ways of thinking in an attempt to create a *radical geography* with the declared aim of curing a sick society.

Harvey (1973, pp15–151) suggests that there are three kinds of theory connected with the potential change or cure of society. (1) The *status quo* theory; (2) the counter-revolutionary theory, and (3) the revolutionary theory. The first seeks to portray as accurately as possible the phenomena it deals with at a particular time, the second appears to be grounded on the reality it seeks to portray, but is designed to frustrate necessary changes, and the third holds out prospects for creating truth rather than finding it. Similarly, Gregory (1978) attacks the positivist way of thinking which he finds to be intimately connected with the quantitative revolution.

We now seem to be in the middle of this new revolution which tries to discard both quantitative geography and positivism, leading research workers to become increasingly afraid of being stigmatised either as poitivists or as quantifiers. We might ask ourselves whether the pendulum has swung too far again in one direction, or whether we are seeing yet another example of the way in which scientific progress takes place through contradictions. Is such a dialectic the only way to progress, or is the dialectic only an educational device with no inherent substance? In dialectics we play with contradictions in order to characterise — and often to caricature — differences between scientists and schools of scientific thought. We are not trying to falsify a viewpoint in order to reject it, but rather to clarify it in order to understand it more deeply. We occasionally, however, forget this objective in the process of emphasising the contrast between opinions in a dialectical approach and we are sometimes misled into thinking that we have arrived at an explanation merely by identifying dichotomies.

Are Dichotomies Overemphasised?

In conclusion, we might point out that it would be foolish to force the dichotomy between the positivist and critical theory approaches to the extent that geography or even individual geographers are characterised either as wholly positivist or as entirely committed to critical theory. The majority of geographers belong neither to one camp nor to the other, for both sets of ideas concern us. The changing balance between them represents variations in the emphasis put on different elements of metatheory. As for 'scientific revolutions' in the sense that Kuhn described them, in geography they have been more like changes in intonation. The basic elements of the old paradigms are not rejected but carried along in some other form. Geography and disciplines like it have not totally rejected old theories in the same way as has happened in theoretical physics (Kuhn's original specialism).

Dichotomies like the one between positivism and critical theory fascinate us because they provide us with simplistic illusions of having understood something significant. When it comes to the point, truth is not simple. James (1972, pp506–7) considered that acceptance of the many dichotomies is a semantic trap and that the attribution of fixed meanings or interpretations to word symbols may result in unreal conflicts between them. He suggested that the following dichotomies have done particular damage to geographical thinking:

(1) that geography must be either idiographic or nomothetic, but not both;
(2) that physical and human geography are clearly differentiated branches of the discipline with separate concepts and methods;
(3) that geography must be either systematic or regional;
(4) that geographical methods must either be inductive or deductive;
(5) that geography must be classified as either a science or an art (the quantitative–qualitative argument).

In fact geography straddles all these dichotomies.

Geography, like may other disciplines, must be both idiographic and nomothetic. There is a continuous transfusion from one to the other: most research contains something of both. An integration of physical and human geography is one of the tasks which justify the existence of the subject. The division between systematic and regional geography is only expedient on strictly practical grounds, one merges into the other. Both induction and deduction should generally be used in the same scientific analysis.

In this sense James argues that much of the antagonism within geography is more of a battle of words than of realities. Since the subject began, every new generation of geographers has asserted that they have founded a new geography, and they have argued for it by overstressing the differences between the 'new' and the 'old.' This is a reason for asking coldly and plainly, 'What is really new?' A study of the development of the discipline shows that it has not experienced any real paradigm shifts. It can turn on a change in intonation, for example, in the direction of more interest than before in the nomothetic aspects, or it can turn towards innovations in technique, method or conceptual thinking. Sometimes a change occurs when a whole new world of information makes it necessary to alter our perceptions, but 'new' concepts are seldom as wholly new as their new discoverers believe, although they can contribute useful additions which increase our understanding of the phenomena we observe in geographical reality. This view of geography is directly opposed to Kuhn's theories, arguing that new knowledge builds up and extends the old. When a new generation maintains that there has been a scientific revolution this is often due to an overemphasised enthusiasm for the excellence of the new material and partially because they have forgotten— or chosen not to notice— what the old tradition stood for.

When we analyse the new 'critical' revolution, it seems to show some of the same characteristics, noted by Taylor (1976) as typical of scientific 'revolutions' in general and the quantitative revolution in particular. The new generation uses the strategy of making the new geography seem so difficult to understand that it is unlikely that the 'old guard' will master it. It is quite clear that Gregory's *Ideology, Science and Human Geography* (1978), Entrikin's paper on *Contemporary Humanism in Geography* (1976) and Gunnar Olsson's *Birds in Egg* (1975) repel those who are not familiar with their methods of argument and concepts of philosophy in the same way as Haggett's *Locational Analysis in Human Geography*

(1965) put off those who were not familiar with mathematics and statistics.

We ought not to avoid real logical and philosophical difficulties, but neither should we express ourselves in unnecessarily complicated language or symbols so that our message only reaches a part of our potential audience. One of the traditional ideals we have in common with history and other 'arts' subjects is to express ourselves as clearly and simply as possible.

'Let one hundred flowers bloom'

Another tradition, derived from history and other 'arts' subjects, is the *hermeneutic approach* (see p87). History, the analysis of literature, and those parts of geography concerned with past events and distant cultures, have no problem of 'double hermeneutics' because what is being studied cannot be changed. Either it is completed, like a work of art, or the people who are involved cannot be influenced (Papuan head-hunters do not read Dr X's thesis on their habits). Many objects of study which are not liable to change, such as the study of the historical geography of Britain, are of importance to geography. The *verstehen* method may be invaluable in the right context, in just the same way as the *scientific methods* which are derived from the 'geometrical sciences' are appropriate to other investigations. Models and theories developed in diffusion and location studies, for example, provide important measuring rods for an improved understanding of geographical reality. We should also realise, when studying contemporary situations which are subject to change, that the results of our research may influence the development of society. Such cases involve the problems of *double hermeneutics*: what we reveal is only what we have understood or modelled. Those concerned, however, may act upon a very limited and crude understanding of our explanations and thus change the situation which is being explained, making new scientific explanations possible. This process can continue in a never-ending hermeneutic spiral. As research workers, geographers also belong to the society under research and are incorporated within its general understanding, beliefs and way of life.

Each of these three approaches and kinds of knowledge: the empirical analytic derived from the natural sciences; the historical hermeneutic derived from arts and history; and the double hermeneutic derived from critical theory as explained by Jürgen Habermas *et al*, are relevant to geography. The extent to which we rely on each of them should be related to the nature of our research problem rather than to strong dogmatic prejudices of our own.

We should approach the problem as to whether the history of geography has been influenced by idealistic or materialistic considerations (see above p73) in the same way. It must be obvious from the comments above that, in the author's view, materialist considerations, and also the nature of contemporary society or 'spirit of the age' (*Zeitgeist*), influence science to a very large degree. These influences are easiest to demonstrate, however, during the earliest phases of scientific development, from antiquity through the middle ages to the nineteenth century. The breakthrough of the positivist ideal changed this situation: science attained a relative independence which made it much easier for individual scholars to influence research ideas and learning. The contributions of Vidal de la Blache in France, of Schlüter in Germany, of Hägerstrand in Sweden, of Haggett in Britain cannot be regarded as having been conditioned by the material base. The work of Marxist-inspired geographers living in capitalist countries demonstrates the very limited guidance which scientists accept from the community or from the materialist base. The relative freedom of science has never been greater, nor have the opportunities for the individual scientist to contribute to the development of his discipline, although other conditions, including the quantity of publications and the degree of specialisation within science, make it difficult for him to reach the whole of his potential audience. Like all other citizens, however, scientists live in a society which will try to control their activities to some extent, especially in so far as the taxpayers generally supply the research funds. In North America especially, where funds are derived from business sources to a great extent, the research worker may experience a conflict of loyalties if he feels that the big firm (or the government agency) has interests which are incompatible with those of the population as a whole. So, while the individual scientist eventually becomes involved in politics like any other citizen, the development of science itself is conditioned both by the genius of individual research workers and by the development of society at large.

CHAPTER FIVE

GEOGRAPHY, A DISCIPLINE OF SYNTHESIS

Scientific Analysis

Before the quantitative revolution, geography was often criticised for being short on theory and long on facts. Even in 1969, Harvey maintained that the commitment of geography to inductive methods 'has not only relegated most geographic thinking and activity simply to the task of ordering and classifying data, but it has restricted our ability to order and classify in any meaningful way. Where explanations have been attempted, they have tended to be *ad hoc* and unsystematic in form' (1969, p79).

Nevertheless, Harvey claimed to identify, both from methodological statements and from empirical studies, six recognisable forms of scientific explanation in geography: cognitive description; morphometric analysis; cause and effect analysis; temporal modes of explanation; functional and ecological analysis; and systems analysis.

Cognitive description is the simple description of what is known, resulting from a more or less successful ordering and classification of the data which have been collected. No theory is involved explicitly but, because the classification usually follows some predetermined ideas about its structure, this involves an element of theory. As an advocate of the 'quantitative revolution,' in 1969 Harvey relegated cognitive description to the lowest order of explanation, although he observed that sophisticated presentations have been made in this way. The advocates of hermeneutic methods in the humanistic schools of thought, on the other hand, often stress cognitive description, maintaining that the quality of an explanation may owe more to the depth of cognition (*verstehen*) to the clarity of expression and perhaps also to the personal commitment of the investigator than to 'technical' methodological procedures.

Morphometric analysis is a special form of cognitive description where systematisation and classification develop from a geometric, spatial, coordinate system. This makes it feasible to undertake network analyses and to study the shape and pattern of the location of towns. Morphometric analysis can lead to certain types of predictive and simulation models. With a knowledge of the geometrical laws of central place theory, the population density and the size and location of two given central places, it is possible to predict the rest of a central place system. Geometrical predictions of this sort have had increasing significance in geography. In morphometric analysis the stress is on *measurement*, whereas studies of landscape morphology (p30–32 ff) usually take the form of cognitive descriptions.

Cause and effect analysis develops from the assumption that previous causes can explain observed phenomena. We look for causal relationships which are, in their simplest form, of the type 'cause A leads to effect B.' This implies that 'cause B cannot lead to result A.' Causal laws may be discovered by the hypothetic–deductive method, or more simply, by comparing data from different phenomena in a region.

After comparing a map showing precipitation on the North American prairies with a map of wheat yields there, we might decide that there is a close relationship between the amount of the precipitation and the size of the crop. We know that precipitation affects the wheat yield but that a high yield of wheat will not bring about a heavy precipitation.

The causal relationship is obviously oversimplified when we only formulate it as one link — we would do better to express the relationship as a chain of causes: the precipitation brings about certain humidity conditions in the soil which will ultimately influence yields of wheat in that area. We cannot even be satisfied with a single chain of causation; the precipitation is not the only significant factor. The nature of the soil, whether sand or clay, for instance, determines the extent to which the precipitation can be stored in those soil horizons where it is accessible to the plants. Other factors include the availability of nourishing salts in the soil, the development and use of improved plant varieties by man, and also fertilisation. We can therefore build up rather complex cause and effect analyses using multiple regression or factor analysis (see Haggett 1965, pp297–303) as tests of the relative weights which should be ascribed to individual factors in the system we are studying.

The general conclusion has been that causality laws are deterministic — that if the cause is present, the effects will follow. If the effect does not follow the cause, the causal law must be rejected, according to this view. The association in geography, during the late nineteenth century after Darwin, between deterministic approaches and causal analyses, brought causal analysis into discredit later on. Many geographers who rejected determinism also held themselves aloof from cause and effect analysis to a large extent.

Since World War II many students have come to realise that a causal analysis does not necessarily involve deterministic causal explanations. This point is made by Montefiore and Williams (1955) in asserting that determinism is really an act of faith, that causal laws are never absolutely true, and that they cannot be verified (see p47). In so far as 'exceptions prove the rule' we try to recognise the 'law' which has the fewest possible exceptions. In this connection 'determinism' is only a sightline for scientific work. The objective is to perfect each causal law as far as possible in order to maximise the *probability* that a given cause will lead to given effects. In recent years, calculations of probability have been included in causal analyses to an increasing extent.

Cause–effect arguments are useful for the analysis of geographical problems. We can use them to analyse simple phenomena, to recognise regular associations, to construct theories, to formulate laws, and so on, but causal analysis has its limitations. It can be employed with a high degree of success in a large number of situations, but is not the only basis for scientific analysis and explanation. The view that the world is directed by cause–effect laws is based on a belief-postulate combining a theory of the 'real nature' of things and a research method which helps us to understand *one* aspect of reality. Causal analyses give us only one of several possible research tools, and cause–effect laws only one form of explanation of reality.

Temporal (or *historical*) *analyses* provide scientific procedures for describing or explaining phenomena in relation to their development over time. From one point of view, temporal analyses may be regarded as forms of causal analysis. History can be seen as a causal series which started at the vaguely defined 'dawn of history' and ends today. In practice it will never be possible to understand such a comprehensive causal series; the analysis must be restricted to some determined period of time. There have certainly been studies which sought to analyse a long process of historical development, but such explanations cannot be so precise as one would normally wish from cause–effect analysis.

Temporal analyses may be classifed according to the assumptions of the research worker in the following ways.

(1) The investigator may assume that there is no provable mechanism which governs development. This has been the most common approach among historians and geographers whereby historical and geographical phenomena are regarded as unique. From such a viewpoint, the analysis becomes a description which does not try for formulate laws for development, but may, however, provide a degree of explanation. Some situations are seen as particularly important and are discussed; others are excluded because they are regarded as insignificant. In addition, there are special but not general explanations for individual phenomena.

(2) The investigator may hypothesise that the observed development is governed by some mechanism. This hypothesis may be that time itself, combined with certain natural laws, is a governing factor. Such approaches are found in W M Davis' cyclical system for the development of landforms through stages of youth, maturity and old age, and in Walt Whitman Rostow's stages of economic development. The problem with these theories has been to relate the situation in an individual region to the relevant stages of development unambiguously. Rostow (1960) points out that some countries may find themselves in two different stages of economic development at the same time. Another hypothesis suggests that development is governed by determined historical laws. The Marxist theory of the economic development of society is a classic example of this approach, often associated with the view that the nature of development is predetermined. Given circumstances necessitate a particular development. Amongst geographers, Ellsworth Huntington suggested a corresponding approach. With other determinists, he believed that men are bound to follow the 'law of nature' and to adapt social forms to 'natural' designs. However, it is quite feasible to set up historical laws without making them deterministic, for calculations of probability may be used to formulate historical probability laws. The problem is that a satisfactory calculation of probability depends on the analysis of large numbers of similar individual occurrences and that it may be difficult for the geographer who is studying a process in a given region to accumulate an adequate number of such occurrences.

(3) Research workers may believe they have sufficient empirical data to state the existence of a form for the mechanism which governs development. In this way, single phenomena can be explained by a satisfactorily recognised process. There is no question of hypothesis formulations, we may describe a situation in terms of a law which states that an event is

probably due to certain previous causes. The natural sciences have been able to establish such probability laws. Our knowledge of biological evolution, for example, is no longer just a hypothesis but is founded, to a considerable extent, on the empirical results of research in genetics and the study of fossil remains of earlier life forms.

Although in-depth studies by geomorphologists in recent years have brought about a fairly comprehensive understanding of the processes which govern the formation of landscapes, human geographers have made little progress in the formulation of laws which govern processes: we would be wise to think in terms of hypotheses and theories in this area of the subject.

Although it was not in line with the traditions laid down by Kant, Hettner and Hartshorne to employ temporal analysis in geography, such analyses have been common. We have seen above how developmental perspectives have been especially prominent in geomorphology, and also in the French school of regional geography. Individual workers have, in fact, adopted a point of view which might be called 'historicism,' believing that the nature of a phenomenon can be *entirely* understood by reference to its development and that complete knowledge can be attained solely by the use of this tool (Harvey 1969, p416; Schaefer 1953, p237).

Such a viewpoint is clearly wrong, for what has been said about causal analysis is equally valid for temporal analysis. The developmental approach gives us only one of many points of view and one of several lines of research.

Both these research approaches have and will be used in geography but changes in the philosophy of the subject during its long history have, to a large extent, regulated their employment. Some schools of thought have wholly rejected one, the other or both of these explanatory forms; other schools have devoted themselves to causal or to temporal analyses. Functional–ecological analyses and systems analyses have been more closely associated with the development of geography and will therefore be treated in more detail.

The Geographical Synthesis

In any discussion of methods, we are concerned with the logic of explanatory models, considering whether they are really tenable scientifically and if there is an in-built logical cohesion in our methods. When discussing the philosophy of the discipline, we are concerned with value judgments, including the metaphysical values upon which we may build a paradigm for the subject. We have seen in practice, however, that fundamental methods have been developed on the basis of metaphysical comprehension. Considerations of method, world outlook and the philosophy of the subject are so involved with each other that it is almost impossible to separate them.

We have already seen that causal analyses were chosen because the research workers concerned *believed* that all phenomena they encountered were determined by previous causes. In the same way research workers retained their interest in temporal analysis because they believed that phenomena could be fully understood as a result of development or process. We would not be surprised to find that the connection between philosophy or belief and method was very close before Darwin's time, but, to some extent, it still exists today. Many research workers do not choose their method of analysis from what they see as appropriate to the task in hand, but prefer what they believe to be the 'correct' procedures.

The view that geography is a synthesising subject has always been basic to the philosophy of the discipline. With the exception of Georg Gerland (p23) and a few others who wished to divide the subject up into a range of (separate) disciplines, a belief in synthesis has been the teleology of geography, the purpose that justified the activities of geographers. However, criticism of the traditional regional synthesis developed alongside the quantitative revolution using the argument that methods which were developed in French regional geography cannot be applied to modern industrial communities (Wrigley 1965). Harvey (1969, p71) maintained that the regional synthesis had been saddled with an unobtainable goal, whilst geographers were in fact working on systematic aspects of problems. Nevertheless, Dickinson (1970, p49) surely exaggerates when he maintains that quantitative geography 'does not concern itself with, and in America directly opposes, integrated studies of chorological phenomena in special regions.'

Modern quantitative or critical geography has not in fact abandoned the idea of synthesis. Many of its leading research workers have been specifically concerned in a search for synthesis through new forms of analyses. Haggett called his general textbook (1979) *Geography — A Modern Synthesis*. One of the more prominent American 'critical' geographers, William Bunge, said (1973, p329) that geography is 'the integrating science, so we call upon co-workers in geology, sociology and so forth to discuss planning for a region or even the lesser labour of just understanding a region with no ambitions humanly to improve it.'

Asbjörn Aase (1970, p13) questions whether geographers will not soon be required to resume discussion of the development of a new regional geography with new methods of attack. He points out that the planning of a modern society requires a good deal of research on the development of syntheses, but that this work at the moment is almost wholly in the hands of economists and architects, leaving geographers on the sidelines in their exclusive concern for systematic geography. Modern geography can, however, develop methods which can be used in the formation of regional syntheses.

In modern geography the argument for the further development of a geographical synthesis is practical in character and goes along the following lines: society requires syntheses at different geographical levels. Geography has worked with practically all the relevant phenomena during its historical development and has devised methods which are central to the formation of syntheses. The syntheses needed, especially in planning, will provide a basis for the development of the discipline and possibilities for new jobs for geographers.

Traditionally, however, the concept of synthesis has been associated with basic philosophical beliefs within the discipline. This is shown in figure 3, which relates modern systems analysis back to the idea of regional syntheses and Ritter's concept of *Ganzheit* (wholeness). The vision of the great overarching unity of nature, held by both Ritter and Humboldt, was characteristic of the idealistic philosophical ideas of their time. All phenomena are related and have a role to play in this unity. Also, the 'wholeness,' in addition to its linking functions, is something more than the sum of the parts — it is seen as an organism. Other subjects study single phenomena but geographers should try to understand the synthesis which reveals the 'wholeness.' The idea of a synthesis is here related directly to philosophical ideas of a particular age.

Functionalism and Functional Explanation

It is basic to this view that scientific analysis and understanding must be teleological — must analyse and explain individual phenomena in relation to their assumed purpose. A teleological explanation relates to the purpose of a phenomenon; a functional explanation relates to its function. Carl Hempel (1959) regards *purpose* as a wider concept than *function*, he considers therefore that functional analysis and explanation is contained within teleology. Other philosophers of science consider that Hempel gives too restricted a connotation to the concept of function. We shall restrict ourselves here to suggesting that teleological and functional forms of explanation are closely related to each other. What interests us is the close association between the traditional view of geography as a discipline seeking synthesis and the functional–teleological analyses and forms of explanation. It has been regarded as logically sound to make the synthesis, or 'wholeness,' our goal, so that each phenomenon should be analysed with a view to explaining its purpose or function within the whole.

It is remarkable that in the 1870s and 1880s, when ideas of geography as a synthetic discipline were weakest, there was little development of functional analysis. True enough Ratzel, like other determinists, did not reject the concept of 'wholeness,' he even discussed the state as an *organism* attached to the land. The determinists, however, were more concerned to advance hypotheses on the laws governing organisms, which they considered to be causes of individual phenomena, than to study the function of the individual phenomenon within the organism or 'wholeness.'

Early in the twentieth century there was a reaction in many disciplines against the simple and deterministic cause–effect arguments and forms of analysis which had characterised the nineteenth century. This welled up into what might be called a *functionalist philosophy*, which sought to replace expressions of cause and effect with expressions which emphasised associations. It tried to develop functional analysis and explanatory forms to replace the *mechanical explanations* which were so characteristic of physics. The term 'mechanic' (from the Greek *mekhane* — arrangement or machine) is used in philosophy to characterise explanatory measures and world views which have used machines as models of both organic and inorganic change. The tendency to prefer functional to mechanical explanations was most marked in biology where attention was directed to complex organisms which must be analysed primarily as 'indivisible wholes.' A flower, for example, can neither be fully understood through an analysis of its stamens, petals, etc, in isolation, nor effectively studied only in its individual 'wholeness' but must also be related to the ecological environment within which it is growing.

There appears to be a range of phenomena which can best be described and analysed with reference to a 'unity' or to a 'system.' It is not because such a 'unity' or 'system' is necessarily governed by an overriding or predetermined purpose, but because individual phenomena must be understood in the light of functional associations and circular causations within the 'wholeness.'

During the first part of the twentieth century, functionalism characterised the outlook not only of biologists but also of many social anthropologists, notably Bronislaw Kaspar Malinowski (1884–1942), and sociologists. Functionalism also affected psychology and economics to some extent, eventually achieving a considerable penetration into those sciences.

Functionalism affected much geographical research in the late nineteenth and early twentieth centuries and philosophy was strongly interwoven with method. Wrigley (1965, p15), points out that some French regional geographers, notably Jean Brunhes (1869–1930), were markedly influenced by functional socio-anthropologists. Just as Malinowski regarded culture as an 'indivisible wholeness' from which an explanation for single occurrences might be derived, the individual region provided a 'unity' for the French regional school of geography. The region was considered to be a functional unit— an 'organism' which was more than the sum of its parts. Hettner and Hartshorne shared this philosophical viewpoint. Alongside his maintenance of the belief that the region is a functional unity which gives a 'wholeness,' Hartshorne (1950) emphasised the need for a functional approach to political geography. Here we can see a link with Ratzel's views (1897) in discussing the *raison d'être* of the state in terms of its 'cohesive geographical forces.'

In political science this was developed by Talcott Parsons in the 1950s and 1960s as the *structural–functionalist* approach, which is closely connected with systems analysis (see p127). A function in this connection is defined as an activity performed by a structure which maintains a system of which it is a part (a definition which requires a deeper understanding of system analysis to become intelligible). The essence of structural–functionalism 'is the system-maintaining activity and the functional approach allows widely differing societies to be analysed because it emphasises their basic functional characteristics which in turn, it is argued, are reflected in deep-seated relatively permanent structural characteristics' (Morgan 1975, p291). The important point here is that the Parsons model has an implicit assumption of an equilibrium-seeking social system which contrasts it with the dynamic model of Karl Marx.

As we will demonstrate (p133), the methodological difficulties with both functionalism and systems analysis begin when we want to analyse influences which change the structure of the system. The Norwegian political scientist Stein Rokkan (1970) has, however, developed Parsons' functionalist framework further to cope specially with the dynamics of the nation-building process, and so has demonstrated that functionalism and systems analysis are not necessarily connected with equilibrium rather than change.

At about the same time as the 'quantitative revolution,' geographers began to distinguish between their basic philosophical beliefs and their use of methods which corresponded with them. David Stoddart (1967, p519) pointed out, in reference to functionalism, that the theme of organic analogies had once been very well developed in biology and philosophy, but that it had long since been discarded there. Harvey also rejected the functionalist philosophy. In maintaining (1969, p445) however, that 'to attack functionalism as a philosophy is not, therefore, to attack it as a methodology', he also suggested that essential insight is achieved by establishing hypotheses of functional connections. The methodological strength of functionalism lies in its support of reciprocal relationships, conditions and feedback in complex systems or organisational structures.

On this basis all forms of scientific analysis and explanation may be used. Cause–effect relationships, temporal, functional or teleological explanations all provide insights into truth, but none of them gives us the whole truth. This conclusion amounts to a break with the creeds of the philosophy of science and it may also lead us to regard the paradigm shifts of Kuhn as a *recessive* stage in scientific thinking as the choice of paradigm was earlier (pp41–42 ff) described as the acceptance of a creed, the choice of a philosophy and an attitude to research which we think that our subject or discipline should generally follow.

Throughout the long history of geography, many basic elements in its thinking have been relatively stable. There has always been a view of geography as a discipline of synthesis. *Holism* has been there the whole time, even if it has changed its character rather markedly. Once it signified some indefinable gift of God; now we talk of it in a methodological way as a functional system which can be subjected to systems analysis.

Before we look further at the most recent developments in geographical method, including systems analysis, we must inspect some of the divisions and difficulties of functional analysis. This is because an evaluation of functional analysis forms a necessary basis for the development of systems analysis.

The Use of Functional Analysis

Functional explanatory models are very common in geography. New York can be explained in terms of its functions as the chief port on the eastern US seaboard, the chief financial centre of North America, and so on. Smaller towns can be 'explained' in terms of their function in a central place hierarchy. Such explanations are usually wholly verbal; they represent a form of description but do not produce any more firm evidence to support an argument.

In order to demonstrate how a functional analysis operates, Harvey takes a central place or market which served the needs of an economy for the efficient exchange of goods and services. Is this explanation sufficient? Central places may *seem* to exist for exchange purposes but may *really* satisfy men's needs to meet each other; the trade function is *manifest* but the social function is *latent*. How far, by asking which are the latent functions and which are the manifest functions, can we develop distinctive kinds of explanation?

In so far as goods and services can only be exchanged at central places, and that such exchange is a necessary part of the economy under investigation, the difference between manifest and latent functions is of no particular explanatory interest. Central places can be explained in accordance with their manifest function. Such a simple explanation is not satisfactory, however, where there are such functional alternatives as pedlars and tinkers. Where such alternatives exist, the central place may nevertheless be preferred. In such a case it would be reasonable to suggest the latent form as an explanation.

It often happens that there are several candidates for one function. A functional explanation of the location of a town or a factory, for example, is not usually a sufficient explanation. There are almost always other locations which would have been just as satisfactory for the purpose.

In most cases, functional analysis only demonstrates the *necessary* location conditions for a town or factory to flourish. It will seldom indicate which conditions are *sufficient* to make any one location the only possible location. Functional analysis can be useful in enabling us to discern some of the conditions necessary for a phenomenon to function within a given system. We cannot, however, ascertain both the necessary and sufficient conditions for a tenable and unambiguous explanation. There is no reason to overlook the functional elements in our attempts at explanation, but we must appreciate the possibility of logical weaknesses in the method.

Individual theoreticians of science have maintained that functional relations are to a considerable extent tenable only within systems which maintain and regulate themselves. The simplest example of such a system is a water-heating system regulated by a thermostat. The function of the thermostat is to register the variations in temperature and to signal them so that an even temperature may be maintained. If we try to imagine this function in terms of social conditions, we would restrict functional explanations to relatively static communities, where the functions help to support the inner balance of the system. Now we know that the communities which geographers are most interested in studying have undergone major developments. A location which could have been explained by functional needs a decade ago is maintained today only as a result of *geographical inertia*; Pittsburgh and Lanarkshire can only be explained as steelmaking centres in relation to the industrial location necessities and economic conditions of the nineteenth century. Harvey (1969, p437) maintained that these views on the logic of functional analysis should only be related to the version of functional analysis as defined by Hempel (1959, pp5–10). The latter defined functional analysis narrowly: 'the contribution that some item makes toward the maintenance of some given system.' It is also possible to define function as a mathematical expression between variables or as an *indicator* of their use-value.

Even if the concept of function is interpreted narrowly, functional analysis may be a useful point of departure for the formulation of individual theories and, perhaps more significantly, for research into alternative methods for the study of complicated systems and the structures of organisations. It may, for example, be useful to ask how central places function in an economy because this question raises a range of other questions and draws our attention to the complexity of the systems within which we work. Whether or not we are going to derive anything useful from functional analysis depends on the extent to which we can specify a system clearly. This is why functional analysis has been so ardently discussed by geographers in recent years. Proposals to use systems analysis and therefore to make use of *general systems theory* in geography were put forward in a number of papers in around 1970. Chorley (1973, p162) said that the application of systems analysis 'greeted by some as a conceptual breakthrough and by others as a jargon-ridden statement of the obvious, has at least served to highlight and rationalise some of the important and long-continued methodological difficulties which geographers have faced.' Systems analysis has been used as a methodological tool to build a new form of geographical synthesis. The role of men in the natural ecosystem is again presented as a common object of study for physical and human geography and 'geography as human ecology' has returned as a subject for discussion.

Geography as Human Ecology

While the use of a more formalised systems theory has only been introduced into geography recently, systemic thinking has been of much longer duration. The history of systemic thinking in geography is closely associated with functional analysis and the consideration of regions as complex organisms or unities and with a viewpoint of geography as human ecology. Elements of such systemic thinking may be found in the writings of Ritter, Vidal de la Blache, Brunhes, Sauer and others. Only during the 1970s, however, did systems theory itself come into focus. In the same way as an appreciation of the logic of functional analysis leads towards a formalised *systems concept*, one of the mainstreams in contemporary geographical thought is leading towards systemic thinking. If we are now beginning to regard this way of thinking as being of decisive importance, it is because of the possibility of uniting geographical methods with geographical philosophy.

The most obvious of the antecedents of systemic thinking was the school of thought which, at the beginning of the twentieth century, wished to define geography as *human ecology*. The ecological concepts which we find implicit in the works of Vidal de la Blache are, however, rather different from those enunciated by Harlan Barrows in his presidential address to the Association of American Geographers in 1922 (Barrows 1923), in which he considered that geography should concentrate on the study of man's associations with his natural environment. His main argument was that the subject had diversified too much and that specialised branches like geomorphology, climatology and biogeography should be separated from the subject, and that geography should concentrate on those themes which lead towards synthesis, with an economic regional geography occupying a central place. Apart from this he gave no clear directions about methods of research; for this reason, and because 'natural science' geographers exerted a powerful influence on the discipline at that time, these ideas made little impact in the 1920s. Barrows' paper *Geography as Human Ecology* was, however, often cited later.

The concept of human ecology may need further clarification. *Ecology* (from the Greek *oikos* = home place, *logos* = doctrine) was first used by the German biologist and popular philosopher Ernst Haeckel in 1868. Biologists classify ecology as the study of the relationships between animals, plants and their environments. Ecologists should, from this definition, study the natural relationships whereby particular species of plants and animals are dependent on each other and on the non-organic environment (climate, soil, the chemical composition of atmosphere and water, etc). The aspects of ecology which impinge on man's biological nature and his relationship to the natural environment, should therefore form the subject matter of *human ecology*.

For many years ecology had a somewhat anonymous existence in biological research. By the 1950s, the concept was being taught largely in terms of autoecology and population ecology. *Autoecology* deals with the reaction of individual organisms to those environmental factors such as light, temperature and humidity which are required by every plant in order to survive. *Population ecology* concerns rates of propagation, death rates, the growth and composition of a population, together with the factors which regulate the numbers of individuals of a species that are found within a region. A third subdivision was *synecology,* which treated the development of plant and animal communities in an area; for example, the slow succession in plant community we see when a forest clearing or field is abandoned. At first a few pioneer species invade such areas, but over time the disposition of species gradually changes until the *climax association* is reached. This is the theoretically optimum biological community which can maintain itself at a place unless there is a further change in natural conditions.

During recent decades, however, ecology has become a major interest of biological research and has been a significant spur to the general debate about the protection of nature. The marked advances in research recently made by biologists are largely due to the development of a new direction within ecology — *systems ecology* — which studies the construction and function of the *ecosystem*.

An ecosystem (figure 8) consists of the biological community at a specific place and the environmental physical circumstances which influence and are influenced by that biological community. The ecosystem is not only the sum of flora and fauna, water, air and soil, for the most critical element is the circulation within the system: the solar energy which comes in, the transport of water and gases, the change of inorganic material into organic material, growth and movement. An ecosystem may be defined for any and every geographical study region from a small puddle to the globe itself.

A significant result of systems ecology has been the development of operational models of circulations and relations between the parts (figure 9), including the means of subsistence, within the ecosystem. So much new knowledge has been produced that radical revision of elementary

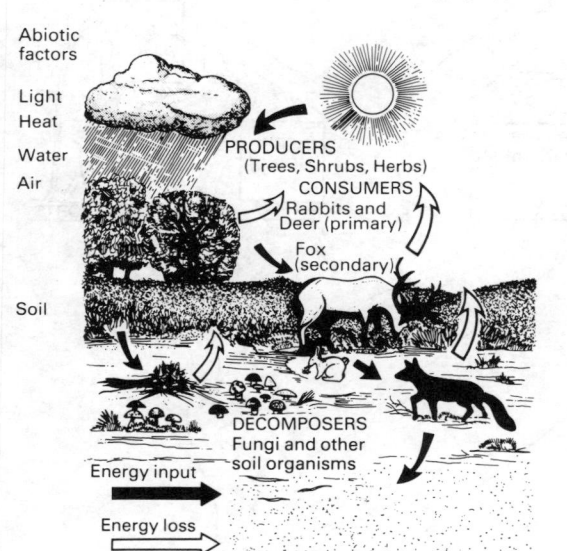

Abiotic
factors

Light
Heat

Water
Air

PRODUCERS
(Trees, Shrubs, Herbs)

CONSUMERS
Rabbits and
Deer (primary)

Fox
(secondary)

Soil

DECOMPOSERS
Fungi and other
soil organisms

Energy input

Energy loss

Figure 8. A generalised ecosystem. (After Bliss et al 1969)

textbooks has become necessary.

To some extent, regional geography can be said to parallel traditional biological ecology, especially population geography and 'synecology.' This is particularly apparent in the landscape geography which is taught through the medium of map interpretation and looks for relationships between the physical and human elements in a region. A collection of papers edited by Robert Eyre and Glanville Jones in 1966 under the title *Geography as Human Ecology* exemplifies this type of study. Another instance is the 'regional ecology school' which developed in Germany after World War II. Stoddart (1967) has little good to say about this form of human ecological analysis. The papers edited by Eyre and Jones 'lack', he says (p522), 'methodological rigour, lean heavily on a largely discarded regional approach, appear to be mainly pedagogical in aim, and ignore developments in ecology itself over the last thirty years.'

My own view is that Stoddart denigrates the role of teaching in the subject too much. Outside the wholly experimental sciences, most of the methods we use in research also have a considerable educational significance — for we use them as a means of presenting our results to our readers (and indeed to ourselves). Hypothetic–deductive methods, for

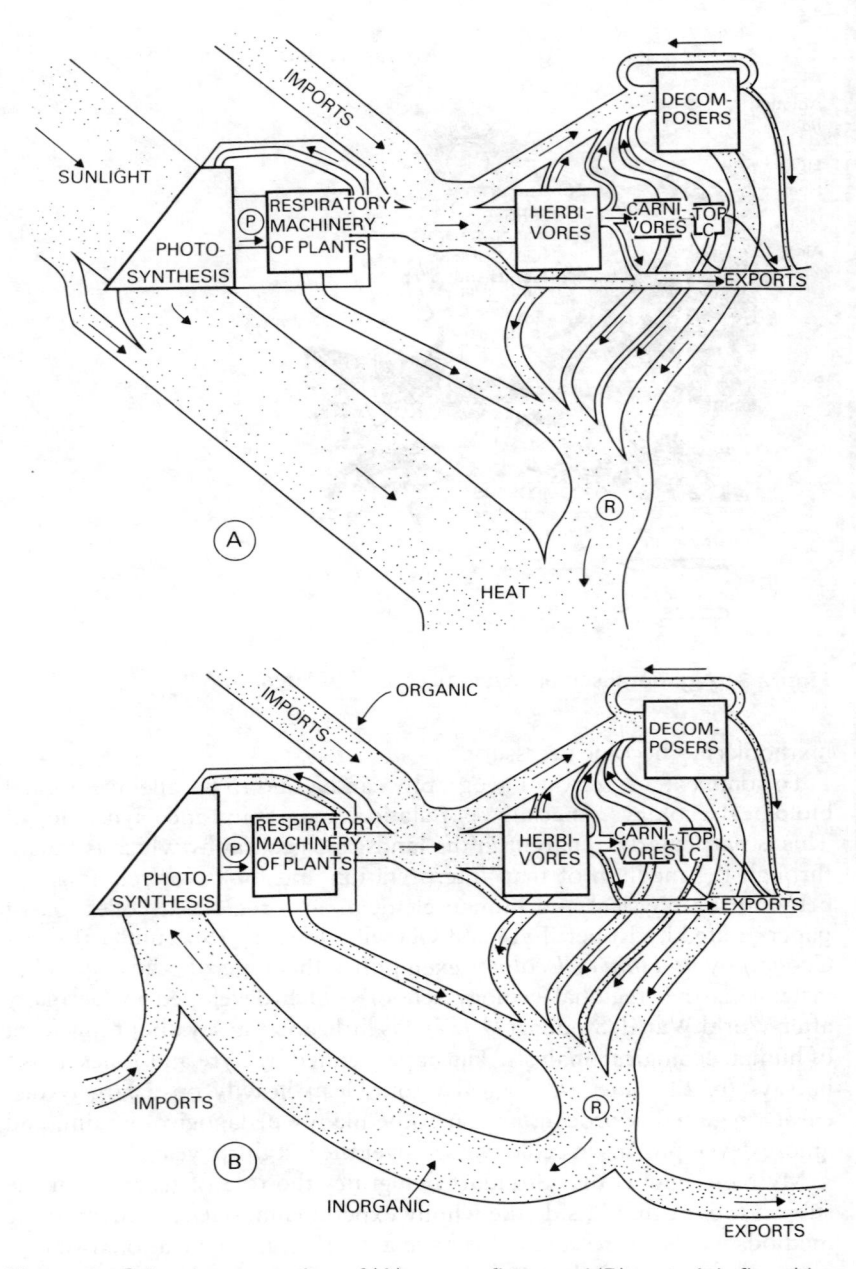

Figure 9. Odum's conception of (A) energy flow, and (B) materials flow (the 'economic cycle') in a simple terrestrial ecosystem. (After Odum 1960, from Stoddart 1967)

example, not only provide a valuable scientific approach to a problem but may also provide a very satisfactory framework for a thesis which can tie the ideas and contents together and, if successfully done, can hold the interest of the reader from beginning to end of a complex story. In the social sciences, hypothetic–deductive procedures are, more often that not, educational tools rather than rigorous scientific methods. Even traditional regional approaches may be useful in expounding geographical knowledge to the public.

If we develop the slogan 'geography as human ecology' we must be aware of the tools of modern biological systems ecology which offer many opportunities for quantitative analysis. Aase (1970, p13) pointed out that 'Individual quantitative techniques offer possibilities for registering variables of a physical geographical nature and of analysing them alongside social variables.' He emphasised the necessity for a grouping of subjects which not only study one or many factors working upon the environment within a regional framework but also have an overall view of both social aspects and the way in which environmental processes operate.

Stoddart shows how the concept of the ecosystem has four elements which interest geographers. Firstly it is monistic: it brings together the worlds of man, animals and plants within a single system wherein the interactions between the components can be analysed, to a large extent, by quantitative methods. Ecosystem analysis, by emphasising the functions and characteristics of whole systems rather than any particular relationships within them, bypasses discussion of determinism and the conflict between physical and human geography.

Secondly, the ecosystem is structured in a more or less orderly, rational and understandable way. A particular advantage is that as fast as the structures of an ecosystem are developed, they can be checked and investigated. Thirdly, ecosystems function: they include a permanent through-put of material and energy. To make a geographical analogy, the system contains not only the perceptible lines of communication but also the goods and people who use them (cf. landscape geography, pp30–1). Fourthly, the ecosystem is a type of general system which uses and can make use of general developed theories of systems analysis.

Stoddart maintains that ecosystem analysis gives the geographer a tool with which he can work (1967, p534), though emphasising that the model-building of the biologists cannot be carried over directly into geography. Chorley considers, therefore, (1973, p157) that 'the ecological model may fail as a supposed key to the general understanding of the relations between modern society and nature, and therefore as a basis for contemporary geographical studies, because it casts social man in too subordinate and ineffectual a role.'

When man began to clear forests and plough the land, the biosphere was changed from a system working in itself and for itself to a resource for one of the species within the system. Man became a conscious manipulator of the ecosystem in order to develop the resources in which he was interested to their maximum use. Chorley (1973, p160) suggests that the ecosystems which biologists study are characterised by negative feedbacks which almost always start to work as soon as something upsets the stability of the system. In the case of a rapid growth in the numbers of an animal species, nature hits back by failing to supply nutrition for the increased numbers, and high death rates intervene until the balance is restored again. The systems in which man is an active participant incorporate stronger positive feedbacks. Men counter nature's rectification by conscious action. When the cultivation of wheat in the drier areas of the North American prairies was about to result in a negative natural feedback, in the form of drought and dustbowl conditions, the farmers, instead of giving up agriculture and allowing the ecosystem to return to its original balance, continued to grow wheat although in strip cultivation every other year. Man very often succeeds in replacing a natural ecosystem by a man-made one. The process of change and of technical and economic development therefore has much greater significance for the analysis of the systems in which men particpate than in the systems which are analysed by biologists.

The ecosystem model has balance, equilibrium, cycling and stability as its basic elements. Chorley (1973, p161) considers that one of the fundamental difficulties facing the development of geographical research today is that geographers may become so involved in such concepts as to overlook the positive possibilities for the manipulation of resources and ecosystems. He refers to Wrigley's characterisation of the regional method of Vidal de la Blache (see p28) and maintains that the Industrial Revolution has made the ecological model inappropriate as a basis for the study of modern industrial communities.

The ecosystem model, Chorley maintains, will be of geographical significance in so far as men can be said to function in the same way as other species. It loses significance to the extent that men succeed in manipulating nature. The use of the ecosystem concept in geographical analysis raises two important questions: are our human systems so much more complex than the biological parts of ecosystems as to limit seriously the geographical use of the model? If so, to what extent does man control the system within which he lives?

To the first of these questions a biologist will definitely answer no, man-made systems are much less complex than natural ecosystems; in his manipulation of nature man always simplifies it. In biological terms, a Mid-West wheat field, because of the dominance of one species, is much simpler than the natural prairie it succeeded, and an important ecological observation is that simple ecosystems are much more liable to negative feedback than complex natural ecosystems. This is also a problem with the town, surely the most manipulated of all ecosystems. Here, simplification creates a lot of dead ends. Sewage, for example, is channelled into what would be called a dead end in a systems analysis, i.e. a limited recipient which cannot transform and recirculate the organic matter, a resource which is thus lost. Most pollution problems are of a similar character, resulting from man's manipulation and simplification of the natural ecosystem, so creating dead ends and ultimately serious negative feedbacks. In this process valuable resources for mankind are depleted. Man may be able to control the system, but if he wants to avoid serious repercussions in the long run, he would be wise to work *with* nature and not against it. The ecosystem approach is of real value in this connection.

This brings us to the current debate about environmental values and the use of resources. Geographers have largely been left out of this debate so far, partly because of the shadow cast by the unproductive conflict between possibilism and determinism during the early twentieth century. I think personally that geographers are so afraid of being branded as determinists that there is little risk that, as Chorley suggests, geographers will become indoctrinated with oversimplified ecosystem models and apply them uncritically to social systems. I believe there is a greater danger that the development of the subject will come to a dead end because we do not dare to use what we can of ecosystem analysis and the concepts which are associated with it. Use of the quantitative models which seemed to be working well in the 1960s may lead us to link ourselves too closely with the models and concepts of economics — a subject which at the moment seems to be confronted with a real paradigm crisis.

The viewpoint expressed above really requires a more comprehensive justification than it is possible to give here. We might, however, suggest that basic theories such as those of John Maynard Keynes, David Ricardo and Adam Smith may no longer be tenable. Adam Smith and David

Ricardo have been classified as *classical economists*; their ideas have been modernised and revitalised in the twentieth century and reappear as *neo-classical economics*. A major argument of classical economics is that total supply and demand are in equilibrium with one another, and that general overproduction and general unemployment are out of question. Another is that the saving of capital is necessary both for society and for the individual. In the 1930s, however, economic theoreticians failed to understand a world which, despite great productive capacity (and also large-scale unemployment and overproduction), could not satisfy fundamental human wants. Keynes gave the theoretical solution in his *General Theory* (1936) in which he stated, among other things, that disturbances in the economy occur particularly because the recipients of income proceed to save more. If they abstain from consumption the result will be overproduction and unemployment. So saving may be harmful; when an economic depression comes we should stimulate consumption. It is better to build pyramids than to let people remain unemployed (Pen 1958). *Keynesian* and neo-Keynesian economics have predominated over classical and neo-classical ideas since World War II, but today a number of economists have suggested that we need a new Keynes with a new 'general theory.' Ezra J Mishan (1969) and Fred Hirsch (1976) have argued with some force that the connection between the growth of welfare and the growth of the gross national product per capita is at least dubious and may not even exist in technically advanced societies. Since Keynes published his celebrated *General Theory of Employment, Interest and Money* (1936), very many economists have worked on the task of operationalising and applying models and theories which aim to create stable economic growth. From the 1930s to the 1960s it was generally held that such stable economic growth was the most crucial interest of the general public, but since the end of the 1960s serious doubts have arisen. Another issue which creates serious problems in economic modelling is that of renewable as opposed to non-renewable resources. It is not possible to set a price on the value of non-renewable resources for future generations. The conjunction of rapid inflation and high unemployment since 1974 is not easily explicable in terms of current economic theories, and so society fumbles blindly after effective remedies.

Chisholm (1975, p52) points out that the trouble with economics is that many of its theories, though logically consistent, are difficult to verify. The field of economics is littered with elegantly and logically constructed but non-operative theories of economic growth and development. Similar comments might be made about many theories which geographers derived from economics in the wake of the quantitative revolution. We may note 'the inadequacies of location theory and the unresolved dispute as to whether urban hierarchies should be based on a $K = 3$, $K = 4$, or $K = 7$ system' (Chisholm 1975, p51). The danger with economics and also with geographical research which has been modelled on economic theory is that in order to achieve intellectual rigour we have to assume away many inconvenient complications. Then we suddenly realise that these complications may be the main issues. Some of the complications might be analysed better through ecological models, although it is quite clear that these also have their limitations. It is important for us as geographers, that our discipline, with its traditional links with both natural and social sciences, is in a rather good position to exploit the interesting and important themes of conflict between what is ecologically desirable and what is economically advantageous.

Individual geographers have tried to bring theories from ecosystem analysis to bear on problems of the interaction of men with nature. Robert Eyre in his book *The Real Wealth of Nations* (1978) criticises Adam Smith's (1776) view that the wealth of a nation is solely the amount and quality of the labour it possesses and the efficiency with which it sets it to work. Smith's point of view might have been reasonable in 1776, states Eyre, but today the supply of natural resources will increasingly be the crucial problem. So Eyre proceeds in an ambitious attempt to provide a calculation of the natural resource wealth of the Earth and its nations. Ian Simmons provides an understanding of man's effects on the biosphere in his textbook *Biogeography* (1979). In another book (1974, 1981) he presents an overview of the ecosystems of the world ranging from those which are least affected by men to the most urbanised systems. Simmons points out (1974, p35) that both economists and ecologists are putting forward theories which are necessary for the understanding of the relationships between man and nature. It is the concepts of the economist, however, which dominate our outlook today. 'The perspectives of ecology are different from those of economics, for they stress limits rather than continued growth, stability rather than continuous development, and they operate on a different time scale, for the amortization period of capital is replaced by that of the evolution of ecosystems and of organisms. So some reconciliation of the two systems of thought might be held to be desirable in which the findings of one science might be translated with some precision into its impact upon the other, and the values suggested by ecology might become operational

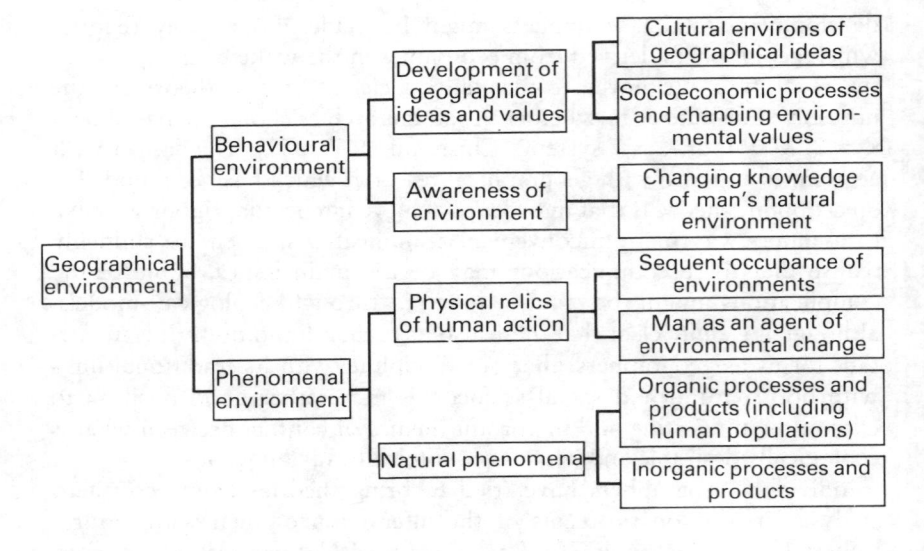

Figure 10. Geography and the environment. (From Kirk 1963)

dicta of economics and vice versa.'

Those who describe geography as human ecology have often defined the concept too narrowly and have presented studies of man's relationship to his environment as if it only encompasses man's relationship to nature and not to his total physical and social environment. For the individual, the environment is much more than just nature. William Kirk (1963, p364) has described the geographical environment in terms which may possibly provide a useful starting point for a discussion of systems in which both ecological and social science theories and concepts may be relevant (figure 10).

Figure 10 lists the factors we may need to consider in such an analysis, but the diagram gives us no guidance in the development of operational models for ecosystems dominated by mankind, or simply for systems where natural, economic and social factors interact. Everyone agrees that all such systems are rather complicated, but in order to obtain a better insight into the behavioural environment of mankind we would need to study *perception*, and this involves an understanding of the research methods of psychology and sociology. We must also find out a little more about systems analysis if we intend to build models of our complicated systems.

Systems Analysis

According to James (1972, p511), a *system* may be defined as 'a whole (a person, a state, a culture, a business firm) which functions as a whole because of the interdependence of its parts.' This sounds like a new version of the previously discussed concepts of the organism and *Ganzheit* (wholeness, totality). The connection is also quite clear in philosophy, where the presentation of a system of internally cross connected parts goes back to the Greek philosophers. Geographers have used forms of systems concepts since the dawn of the subject. Despite its venerability, a systemic approach has tended to remain a philosophical concept rather than providing guidelines for practical research. No methods and techniques had been developed to enable the analysis of complex systems in an accurate way before World War II. Systems concepts were invoked in descriptive contexts with particular reference to consideration of the balance of nature.

The first *general systems theory* was put forward by Ludwig von Bertalanffy, who began his professional career as a biologist during the 1920s. In biology he discovered that colleagues tried to increase their knowledge of the nature of organisms by dissecting them into smaller and smaller parts. It struck him that, until we study the individual organism as a system of multifarious associated parts we would not really understand the laws which govern the life of that organism. After a while, he realised that this idea could be extended to non-biological systems and that these systems had many common characteristics over a range of sciences. It was possible to develop a general systems theory which gave the same analytical framework and procedure for all sciences. A general system is a higher-order generalisation of a multiplicity of systems which individual sciences have recognised. Von Bertalanffy saw general systems theory as a way of uniting the sciences, but when he presented this idea at a philosophical seminar in Chicago in 1937, the academic world was not ready for such a theory.

At that time the tendency was to concentrate on detailed investigation of separate phenomena and scholars were sceptical of general theories. Physics was almost the only science concerned with general theory. The majority of research workers at that time were looking for cause–effect explanations. Since World War II, however, the growing quantity of multidisciplinary research attempts to investigate more complex phenomena, and the development of statistical methods and computer technology, have prepared science for general systems theory.

The development of *cybernetics* (from the Greek *kybernete* — helmsman) is particularly relevant here. This new branch of science was founded in 1949 by a group of American scholars led by the mathematician and physicist Norbert Wiener, and may be defined as the study of regulating and self-regulating mechanisms in nature and technology. A *regulatory mechanism* follows a programme, a prescribed course of action which produces a predetermined operation. A water tap is a simple regulatory mechanism, and a thermostat is a *self-regulating mechanism* normally used to maintain a predetermined level of operation constant. In nature, there are a very large number of self-regulating mechanisms, such as the automatic regulation of body temperature. Wiener believed that these self-regulating mechanisms follow certain common laws and that they can be described mathematically in the same way. Whilst the regulation is very precise in nature, in human societies it is defective. Wiener considered that cybernetics could also be used in economic and political fields and that only by using cybernetics could mankind achieve good government.

Cybernetics is primarily concerned with the control mechanisms in systems and with communication processes which determine their successful working. It places emphasis on the interaction between components rather than making sharp distinctions between cause and effect. Between two components, causal mechanism may work both ways. An impulse which starts in one part of the system will work its way back to its origin after being transformed through a range of partial processes in other parts of the system. Part of the mathematical basis for cybernetics is found in information theory. Cybernetics is now regarded as a useful discipline for approaching the more philosophical aspects of general systems theory which von Bertalanffy began to explore in papers and books during the 1950s and 1960s, leading to his most-quoted book *General Systems Theory — Foundations, Development, Applications* in 1968. To understand something of his theory we must have a more precise understanding of systems concepts.

The kinds of system which we can analyse are abstractions. Every real system (such as a landscape) is indivisibly complex. A real system is composed of an endless number of variables which different research workers, with different aims, may, with good reason, analyse in different ways. We may form several different abstract systems from one real system. As a method, systems analysis concerns abstraction rather than truth. The system must therefore be seen as a useful abstraction or model which enables a particular form of analysis to be made.

The abstract character of a system is emphasised when we realise that a system, if it is to be analysed, must be 'closed.' An 'open' system interconnects with its surroundings. All real systems (such as landscapes) are open systems. When we analyse a system we can only consider a finite number of elements within the system and the reciprocal relations between them. The elements and connections which we are not able to consider in such an analysis must be disregarded completely. We have to assume that they do not affect the system. In the analysis of a region, we can of course take into account individual influences and single elements which are not geographically located within the predetermined area or region. The abstract system remains closed all the same because we enclose those elements and relationships in our conceptual model. The system is not synonymous with the geographically bounded landscape, but is congruent with the model we have made of it, represented by the elements and connections we have chosen to enclose or consider.

In other words, we can only study a system after we have determined its boundaries. This presents no mathematical problem since the boundaries draw themselves in so far as some elements are defined as belonging to the system and some as lying outside it, although it is not all that easy to choose those elements, in practical geographical research. The demarcation may seem obvious in some cases, as when the system is discrete and has well defined connections with the surrounding environment. These connections can then be built into the model or we may exclude them. As an example of such a system, Harvey (1969, p457) describes a firm which functions within an economy on the basis of a particular set of economic circumstances. When we analyse the *internal* relations and elements *within* the firm as a closed system, we must regard these circumstances as unchangeable. To extend the boundaries of the system so as to include the changing social and political relationships in the society of which the firm is a part may well alter the result of the analysis. So, even in this simple case the drawing of the boundaries creates problems.

The necessity of abstraction and closing also applies to cause and effect analysis and to more theoretical temporal analyses. When we say that A causes B, we have determined that A and B shall be fixed elements which stand in a predetermined relationship to each other and that no other elements will affect the connection. This means that we have determined a closed system around A and B. We can therefore regard a cause–effect analysis as a form of systems analysis where the assumption of a one-way operation severely simplifies the system and its reciprocating activities.

By identifying the set of *elements* which we believe best describes the real system we can construct an abstract system in order to model a real situation. For example, in a large industrial company engaged in several branches of activity, the head office and each of the branch offices form its constituent elements. Mathematically expressed, the system consists of a set of elements $A = (a_1, a_2 \ldots a_n)$. To this expression should be added an element a_0 which represents the environment of the system. In this case, a_0 can be the economic system within which the firm operates. We can then infer a new set of elements $B = (a_0, a_1, a_2 \ldots a_n)$, which includes all the elements in the system plus an extra element which represents the environment. We can then investigate the *connections* between these elements. Analysing the company, we can see whether there are any connections between the branches, and, if so, between which branches, or whether there is only direct contact between the head office and the individual branches. We can observe whether the contacts go both ways and what the contact model implies. Let r_{ij} represent the connection between an element a_j and another a_i. If $r_{ij} = 0$, there is no direct connection between element a_i and element a_j. In our example this would imply that a particular branch has no contact with another branch. All these connections can be expressed as a new set $R = (r_{0n}, r_{02} \ldots r_{0n}, r_{12}, r_{13} \ldots r_{(n-1)n})$. A system can therefore be defined in the following terms: every set $S = (A, R)$ is a system. In general, therefore, a system consists of

(a) a set of fixed elements with variable characteristics;
(b) a set of connections between the elements in the system;
(c) a set of connections $(r_{01} \ldots r_{0n})$ between the elements in the system and its environment.

We can study three basic aspects of every system: structure; function; development. The structure is the sum of the elements and the connections between them. Function concerns the flows (exchange relationships) which occupy the connections. Development represents the changes in both structure and function which may take place over time.

Elements are the basic units in the *structure* of a system. From a mathematical point of view an element, like a point in geometry, has no definition. A mathematical analysis of a system can therefore proceed without further consideration of the nature of elements, but we need to conceptualise the phenomena in such a way that they can be handled like elements in mathematical analysis. There are two fundamental problems here. The first is a scale problem; an economic system, for instance, may be said to consist of firms and organisations. Organisations such as trade

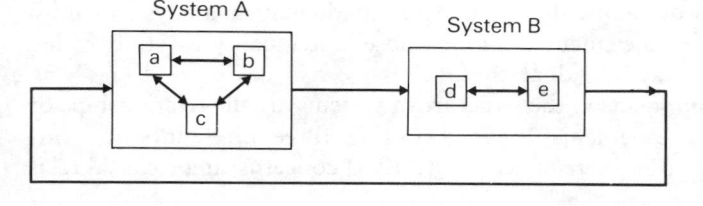

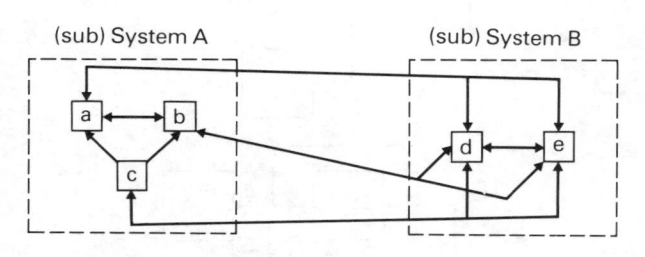

Figure 11. Systems and subsystems. The upper diagram shows system A and system B interacting as units, with smaller system interactions going on within each system. The lower diagram shows systems A and B interacting at lower levels. (From Blalock and Blalock 1959)

unions are also themselves systems which include branches and groups of workers on the shop floor. The latter group is also a system of individuals each of whom may be visualised as a biological system and so on. Whatever we choose to regard as an element at a particular level of analysis, may also be a system at a lower level of analysis. This brings a host of problems with it. We can, for example, regard an element as an indivisible unit (like a branch in a firm resolving or acting on a resolution). A branch can also be seen as a loose association of lower-order elements (individuals in the branch who have contacts with other individuals in other branches). These two points of view are developed in figure 11.

After we have decided which scale to use, another problem in system building is how to identify the elements. Identification is particularly difficult when we are dealing with phenomena which have continuous distributions, as, for example, when precipitation forms an element in a system. Identification is easiest with elements which are clearly separated, such as farms. In mathematical systems theory, single elements are variables. In order to use the apparatus of mathematical analysis we must often formulate an element as a unit whose individuality we study rather than define it as an individual itself.

Further components in the structure of systems are the relationships or links between the elements. We may consider three different basic forms of relationship: a *series relation* (figure 12.1) concerns single causal rela-

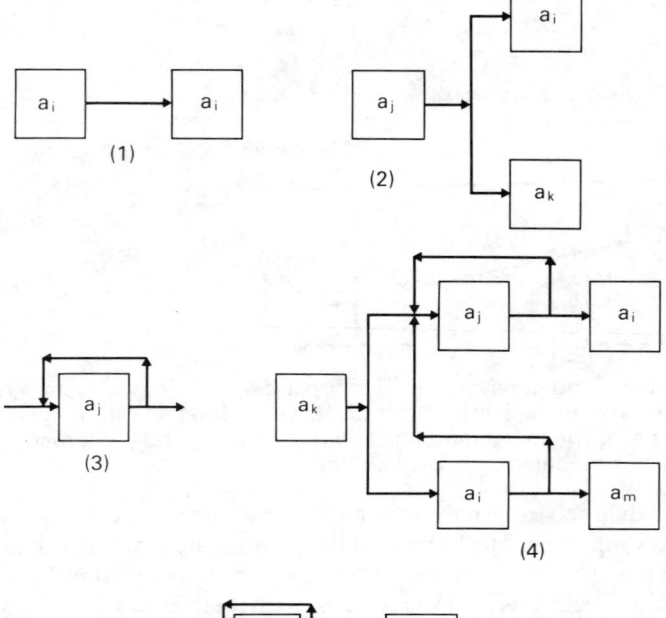

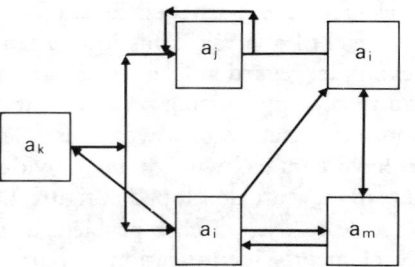

Figure 12. Relations between elements in systems. (1) Series relation; (2) parallel relation; (3) feedback relation; (4) simple compound relation; (5) complex compound relation. (Adapted from Harvey 1969)

tionships which may be linked in causal orders. A *parallel relation* (figure 12.2) occurs when two or more elements affect a third or, inversely, when one element affects two or more others. Such relationships are found in the more complicated cause–effect systems (p106). A *negative feedback relationship* (figure 12.3) describes a situation where an element eventually influences itself. These three basic forms of relationship can be linked in a variety of ways in order to construct complicated systems (figures 12.4,5).

The *function* of a system is concerned with the flows, influences and reactions within the network provided by its structure. Analyses of functions are concerned with the internal flow in the system such as the energy flows and food chains in an ecosystem. If we only want to know which function the element performs in the maintenance of the system, we can carry out a functional analysis similar to that discussed above. We may wish to go further and to discover a mathematical or numerical expression for the functional associations between the elements. It may be possible to represent the association by, for example, regression lines. When we are dealing with flows which may be directly quantified, such as energy flows within an ecosystem, and telephone contacts within a firm, it is fairly simple to display them diagrammatically by drawing connecting lines between the elements and illustrating flow by varying widths of arrows (figure 9).

The *development* of a system primarily involves the influences which come in from the environment and affect the elements. If any change takes place in the environmental circumstances, it will affect at least one element in the system at first. An impulse will then be carried through the whole system until all the associated elements are affected. This may eventually result in temporary or permanent changes in the flows and also in the functional relationships within the system. Influences which change the functional relationships within a system, but not the structure of the system, are most appropriately analysed through systems analysis. An example of this is input–output analysis as used in economics.

Our problems with systems analysis begin when we want to analyse influences which change the structure of the system. As Chorley (1973, p166) points out, it is much easier to construct equilibrium models than models which are changed over time. Most of the theories which are developed for systems analysis concern what can be called *static* or *adaptive systems*. Static (homeostatic) systems are defined by Harvey (1969, p460) as systems which resist any alteration in environmental conditions and exhibit a gradual return to equilibrium or steady-state behaviour after any change which has affected them. An adaptive system shares many common characteristics with homeostatic systems, but differs to the extent that it will try to alter its state towards a 'preferred state' if it is not already in that state. Harvey (1969, p461) suggests that these developments are most characteristic of systems which have normally been regarded as goal-seeking. As a concrete example he suggests that a preferred situation for a journey-to-work system is when the demand for workers by a factory can be met from the working population of the neighbouring housing estates. 'If a rehousing project takes residences further away from an employment opportunity then the system adapts by altering the parameter of the distance function.' The system adapts itself to the new situation but without undergoing any real change in its structure. The theory of what Harvey calls dynamic and controlled systems is less well developed. In *dynamic systems* a feedback mechanism results in the system changing itself through a series of unrepeated states. Feedback can, for example, lead to a situation where new, preferred circumstances are identified. Economic growth models can be seen as such dynamic systems. The problem with such models is, however, that they *only* exhibit the dynamic aspect. Chorley (1973, p164) maintains that real systems are neither in equilibrium nor dynamic, but that they lurch from one non-equilibrium state to another. He suggests that this is especially true of those systems which geographers are most interested in studying, namely the ecosystems of the Earth which man constantly affects and changes. The ecosystems modelled by biologists are, however, to be classified as static (or to some extent adaptive) systems.

Geographers are primarily interested in studying systems whose most important functional variables are spatial circumstances such as location, distance, extent, density per areal unit and so on. A system where one or more of the functionally important variables are spatial may be described as a *geographical system*. Although the majority of ecosystem models do not contain spatial variables, it is relatively simple to build such variables into this type of model. This signifies that geographical systems must easily allow themselves to be constructed as static or adaptive systems. It is difficult to make a geographical system dynamic for then we must combine time and space in the same model. Space may be expressed in two dimensions by cartographical abstraction, but when we introduce time we need a three-dimensional presentation. We may be able to present a satisfactory explanation for such a system but it is very difficult to handle and analyse. These problems have been thoroughly analysed in the presentation of *time–space models* which have recently been developed in Lund (see Carlstein *et al* 1978).

Some problems may be solved by developing geographical models which may be classified as *controlled systems*, which are defined by Harvey (1969, p462) as systems in which the operator has some control over the inputs. *Systems control theory,* which has developed within cybernetics, is based on such systems. Controlled systems are particularly useful in planning situations when the objective is known and the input in the economic geographic system has been defined. In most cases we can control certain of the inputs, but others are either impossible or too expensive to manipulate. If we wish, for example, to maximise agricultural production, we may be in a position to control the input of artificial fertilisers, but we cannot control the climate. Partially controlled systems are therefore of great interest. In future applied geography might well focus on the development of models for controlled systems and attempt to show how spatial systems can be organised or developed by the manipulation of a few key factors.

Chorley (1973, p166) maintains that 'the kind of geographical methodology which . . . is increasingly necessary is analogous to that used in analysing a man–machine system. In the geographical context the 'machine' is made up of those systems structures of the physical and biological environment which man is increasingly able to manipulate [either advertently or inadvertently].'

Our increased knowledge of environmental conditions, as described in the biologist's ecosystem, leads us to appreciate the extent of the need for the development of planning and control systems. Many of the scientists engaged in research into possible future conditions fear that the positive feedback mechanisms in the form of technological development and control which have led to an exponential increase in population, industrial production, etc, will, in the long run, result in a dramatic crisis of pollution, hunger and shortage of resources. One of the causes of such a crisis would be the long-term suppression of natural negative feedback mechanisms (Meadows *et al* 1972).

Chorley suggests, however (1973, p167), that this is unlikely to give rise to a situation where man loses his control over nature. It is important that this challenge be met with new positive feedback in the form of better planning and control — a task for geographers. The challenge is to develop controlled planning systems for future communities using our ecological knowledge as a basic input. We can no longer in practice take the effectiveness of economic development models and technological solution mechanisms for granted.

Edward A Ackerman, who was one of the first geographers to point to the value of systems research, expected 'systems engineering to play an increasingly large role in coping with the social and economic crisis that technological change has brought' (1963, p436). Within ten years of Ackerman's statement, in the early 1970s, many prominent geographers were advocating systems analysis as a major field of activity for the subject, although they were fully aware of some of the problems involved in its use.

The late 1970s saw a more critical appraisal. Chisholm (1975, p36) believed that 'by the end of the present decade it will be generally accepted by geographers that while systems, like regions, provide useful frameworks within which to work, they are all too frequently intangible things that with maddening regularity retreat from the researcher — just as the bag of gold at the rainbow's end eludes the seeker after riches.' The reader may well appreciate this statement after working through the discussion of the theoretical problems of systems above. Systems may seem too abstract; it may seem difficult to get to grips with how to use the theory. If, however, we see how the central ideas of systems analysis have been used for many years in practical research apparently without drawing inspiration from the literature of systems theory or using its associated jargon, the concept may become less mysterious. Most practical work we do in geography may be conceived in terms of systems analysis. Systems analysis may provide a useful systematisation of our models, theories or structured ideas, but it is not necessary to refer to systems analysis and its mathematical implications when we are doing practical research. For instance, a world map of iron ore production and trade may be described in systemic terms: the elements are the producing and the consuming centres, the relations or links are the trading lines, the amount of iron transported along the different lines depicts the function, and maps showing these situations at specific time intervals would describe the development of the system.

Gregory (1978, pp42–7) criticises both systems analysis and general systems theory on the ground that they are intrinsically associated with positivism. The concept of one systems theory which is relevant for all the sciences may be seen as a fruit of the positivist concept of one science, one method. He is further afraid that the prominence given to control systems may lead to instrumentalism (see p98): 'In case this is misunderstood,' he states (1978, p45), 'the sincerity or otherwise of advocates of systems analysis is not at issue; rather, the positive conception of science to which they subscribe *in itself* entails a particular conception of practical life, irrespective of what their own intentions might be, and the connection with the two are determined by what Habermas describes as a deep-seated *cognitive interest* in technical control.' This is clearly a problem because it might be difficult to take care of the individual within the great system. In general we calculate and model average situations and build our control models on this basis, whereas the needs and opportunities of individuals and groups may deviate far from this average. This problem is however tackled in new lines of research within *welfare geography* (p72).

As to the political bias of systems analysis, as indicated by Gregory, we may point to Harvey (1974, p270) who states that 'systems-theoretic formulations are sophisticated enough (in principle) to do everything Marx sought to do except to transform concepts and categories dialectically, and thereby transform the nature of the system from within.'

Description, Analysis, Prognosis

Asbjörn Nordgaard (1972, p29) says that in the development of a science we can distinguish three broad stages: descriptive, analytical and predictive. Description is the first step and the simplest; it is concerned with the description and mapping of phenomena. The analytical stage moves a step further by looking for explanations and seeking the laws which lie behind what has been observed. By the predictive stage, the laws have been studied so thoroughly that we can use models to predict occurrences.

The discussion on systems analysis above has suggested that geographers are now moving into the predictive stage. They are trying to develop models for controlled systems which may be used to guide development in the future. Aase (1970) considers that geographers must involve themselves in the formation of planning syntheses. Chorley (1973) argues that geographers should be concerned with the formation of controlled systems, Berry (1973a) recommends 'process metageography' and the analysis of complex systems for taking decisions about environmental location conditions. Hägerstrand (1973) says that we must try to use our experience of translating the world into geometrical terms in order to develop a time–space geometry with particular emphasis on entering projects with a limited space budget. This may develop a pattern of future-oriented deductive thinking in the subject such as has not been possible before. Garrison (1973, p247) declares:

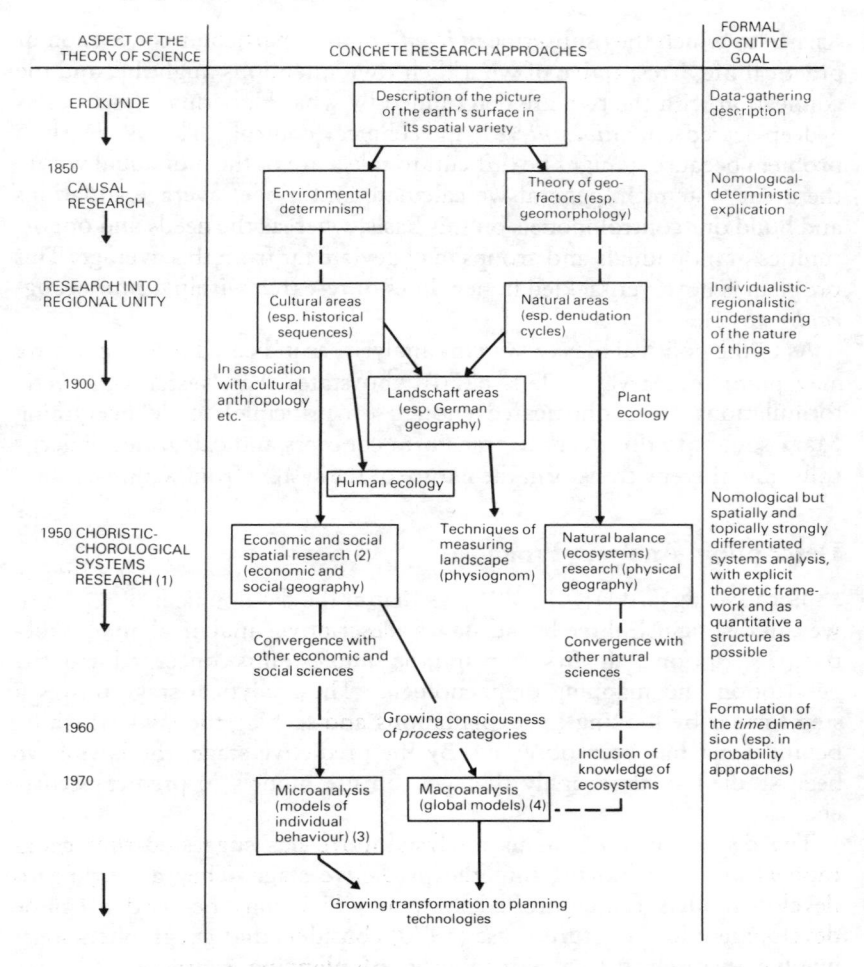

Figure 13. The development of research in geography. (From Bartels 1973)

'While our current concern with systems and relations represents a high level of understanding, it is not nearly high enough. It must extend to the alternative scenarios that these and new relations might follow, and to the problem of choice among these alternative futures.'

Nordgaard maintains that, for geographers, the predictive stage is still far ahead. The nature of the discipline suggests a strong emphasis on reasoned description. Little by little, he hopes our explanations may

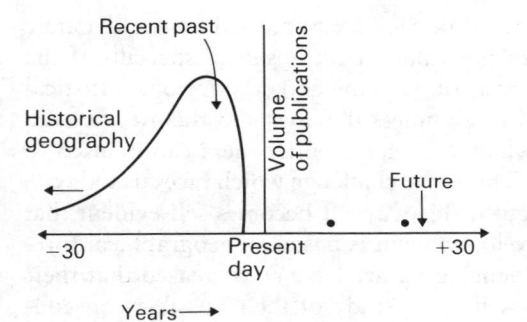

Figure 14. The time distribution of geographical research. Most published geographical research deals with the recent past—very little is concerned with the future. (From Haggett 1979)

become more precise and quantitative — so that the predictive stage draws closer (Nordgaard 1972, p29).

Bartels (1973) presents a schema (figure 13) which, in a rather instructive way, shows the main stages in the development of geographical research from the descriptive through the analytical towards the predictive stage. Figure 13 may usefully be compared with figure 3 (p44). There is no need to add to what has already been said on the development of the subject apart from pointing out that the quantitative 'revolution' turned geography into a predominantly analytical discipline and that this was the most significant aspect of the shift in interest within the subject which took place during the 1950s and 1960s. The predictive stage has hardly arrived yet.

Haggett (1979, p588) points out that the adoption of 'future geography' is something new amongst geographers. If we look at papers in journals, books and maps which were published by geographers in 1979, he says, we find that the majority relate to the world as it was in the early 1970s: the spatial models, ecological relationships and the regional systems which could be analysed at that time. Some of the research workers were involved in describing and interpreting the geography of earlier decades and centuries. Only a few research projects were about the geography of the future. Haggett illustrates this with a simple diagram (figure 14). He explains the lack of treatment of contemporary affairs by the fact that geographers are very dependent on empirical data which have been collected and published by officials. There is a long time gap between the collection and the publication of data; it also takes time to analyse the data and to have the results published. Research reports are therefore beginning to be historical by the time they are published. In our dynamic society this markedly reduces the value of the results, especially if the material presented consists largely of mapping and description. Statistical studies can have only limited value unless the laws and theories derived from them and the models which are suggested by them can be used to develop or govern the future. This is the challenge which faces us today.

Anuchin (1973, pp61–2) put it this way: 'It becomes self-evident that production can only fully develop when it is based on geographical forecasting. Man's means of influencing nature have so increased that their application cannot continue without a study of their possible consequences. Geography, however, has not shown itself ready to solve this problem. The existing main branches (geomorphology, hydrology, demography, economic geography, etc), in spite of their usefulness, are completely unsatisfactory when one is concerned with questions connected with the evolution of the variables involved in regional complexes of the geoenvironment which control the possibilities of the development of production. For this purpose one must have synthetic general geographical studies, the results of which would provide practical forecasts of the consequences of interference with natural processes, which inevitably are taking place. It is necessary to have a science concerned with the utilisation of nature, a science which will connect natural science with the group of social sciences — what is needed is a geography without adjectives.'

Anuchin is largely repeating the old view that geography must be a synthetic science. What is interesting is the justification, for the Soviet geographer is asserting that geography can only add something useful to research on the future if it can make the geographic synthesis work.

A New Synthesis?

Haggett (1972, 1979) has attempted to answer this question. He has tried to develop a new form of synthesis which diverges from the traditional division of the subject (figure 15). He emphasises that the historical divisions are important if only because Universities use them as a basis for their courses. It is more valuable, he thinks, to divide the subject up in relation to the way in which it analyses its problems. The three new main groups are defined as follows.

(1) *Spatial analysis* concerns itself with the variations in the localisation and distribution of a significant phenomenon or group of

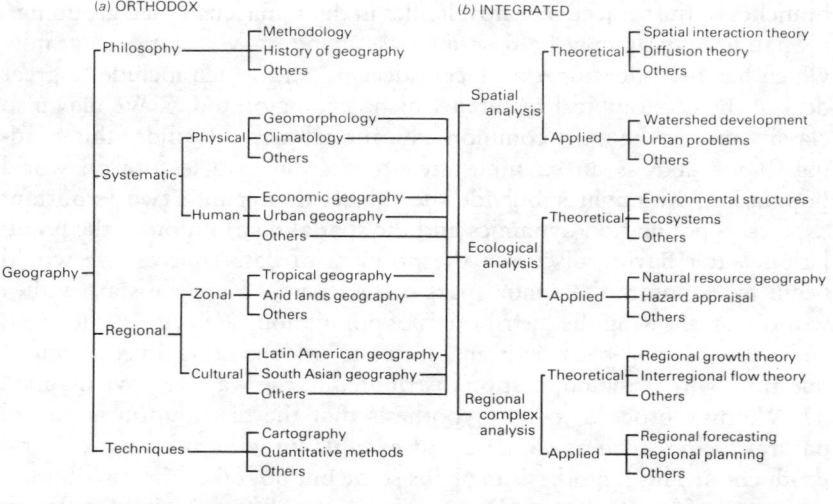

Figure 15. The internal structure of geography. (From Haggett 1979)

phenomena; for instance, the analysis of variations in population density or poverty in rural areas. Which factors control the distribution pattern? How can these patterns be modified so that the distribution becomes more effective or just?

(2) *Ecological analysis* concerns itself with the study of connections between human and environmental variables. In this type of analysis we are studying the relations within particular bounded geographical spaces, rather than the spatial variations between regions.

(3) *Complex regional analysis* combines the results of spatial and ecological analysis. Appropriate regional units are identified by areal differentiations. Connecting lines and flows between the individual regions may then be observed.

This classification may initiate new thinking on the subdivision of geography, although the extent to which it provides the ultimate answer to the old problems of geographical synthesis is more doubtful. In the development of his textbook, Haggett (1972, 1975, 1979) has tried to arrange his themes under the primary headings of ecological and complex regional analysis. His work shows clearly, however, that satisfactory models for complex regional systems, let alone complex ecosystems, have not yet been developed. It is necessary to present models which have been developed through spatial analysis within the traditional systematic branches of the subject. The novelty lies in their unaccustomed grouping.

Spatial analysis may be described as a type of systematic geography which has been developed in recent decades and which includes a great deal of the geographical research which is going on today. We may also classify one of the most common educational methods under this heading. If we take as an example the preparation of a lecture on world population, we might subdivide the subject matter into two important aspects — population dynamics and the spatial distribution of the population. After having discussed the problem of data sources, we would begin to elaborate the spatial part of the theme. Here we start with a world map showing the distribution of population. We begin rather than end with the map because it enables us to ask the crucial geographical question 'why is the population distributed in the way we have mapped it?' We may proceed to the hypothesis that the distribution is due to natural factors such as climatic and edaphic limitations on agricultural production. This hypothesis explains some but not all of the distribution pattern. It fails, for example, to explain the difference in population density between the lowlands of northern China and the North American prairies. We should have to take historical, social, political, economic and a number of other factors into consideration in order to achieve a full explanation of world population distribution. This type of procedure can be used in a number of cases and illustrates what a simple spatial analysis is about.

The definition of ecological analysis is founded on an expanded understanding of the environment as explained above (p126 and figure 10). When we are thinking of simple regions, ecological analysis applies to what have traditionally been called *homogeneous* or *landscape regions*, in which each aspect within the region is similar but differs from the equivalent aspect of other regions. The homogeneous region is also defined by its uniform character. In a physical context the boreal forest, for example, is a homogeneous region but it is always possible to subdivide this region further and it is also rather difficult for scholars to agree on the demarcation of its boundaries.

In education and in practical research, ecological analysis is closely related to chorology and regional geography which seek to understand the region as an entity. There are different ways of presenting as complete a picture of the man–habitat relationship within an area as possible. One is to analyse the landscape with all its human and natural morphological features as they are presented on a map or on an air photograph and try to get at the factors which we see. Another way is to study the processes of change in the region, starting at an appropriate point in time in the past and following the course of the development up to the present, concluding perhaps with our ideas about future scenarios. Both of these simple methods are highly useful in increasing our understanding of current human ecological problems and in our appreciation of the choices for our future development. We may well see a revitalisation of both regional historical geography and landscape geography in the near future.

Complex regional analysis has been chiefly involved with *functional regions*. The functional region is defined by the contact relationships between a centre and its tributary surrounding regions. The boundary of this region is established at the point where the influence of the centre is no stronger than that of another centre. Functional regions are linked, as we know, through complex hierarchical models, so Haggett is using complex regional analysis to unravel these relationships. It is quite clear that complex regional analysis is closely connected with the types of study undertaken by the pioneers of the 'quantitative revolution' like Christaller, but most of this research investigates only limited systematic aspects of the complex regional picture and has not yet reached the point of studying total complexes. The field of enquiry of modern quantitative geography is however by no means limited to complex regional analysis, it is also active in the other fields of analysis.

Modern geography can be said to focus its attention on *spatial analysis*, a systematic geography constructed with newer, expanded models; *ecological analysis*, a regional geography based on homogeneous regions; and *regional complex analysis*, a regional geography based on functional regions. Although this simple division is based on concepts found in an introductory textbook, it may have some importance for the further development of geography as a science devoted to synthesis.

From Orthodoxy to Pluralism

Whether geography will ultimately provide the grand synthesis is, however, seriously doubted by a number of writers. Gregory (1978, pp170–1), for instance, considers that the idea that geography might develop as a bridge between the natural and social sciences has been advanced more as a pious hope or rueful excuse than as a serious proposition. One reason why geography has not developed this synthesis very far is that the natural and the social sciences keep pulling it in different directions. Hannerberg (1961, 1968) argues that the systematic branches of geography are separate sciences in their own right; it only creates logical difficulties if we regard them as integrated parts of a synthetic discipline, geography. He states that the idea of geography as a discipline of synthesis is so general as to have no meaning in concrete scientific work.

The broad field of enquiry traditionally attributed to geography has required that research workers in the subject deploy a wide variety of skills and also that research workers with experience in mathematics, statistics, biology and geology, as well as in history, sociology and economics should be recruited. As no single individual can cover more than a couple of these fields it has been necessary to build up a staff of specialists in each of the branches of geography if the whole synthetic discipline is to be presented to students. The saying that 'geographers specialise in not being specialists' in no way applies to the staff of university departments of geography. Research workers must specialise in order to create something worthwhile; in the normal course of events an individual will work in a field that interests him and in which he has a fair amount of background knowledge. As specialisation develops more and more, individuals will resent the imposition of any kind of paradigm, even if it only states the aim of geography as being a discipline of synthesis.

James Bird (1979, p118) observes that there 'certainly are basic strains within geography, and if one paradigm is plastered across the subject, it will soon be broken by the disjunctions below.' It is, he believes, a hopeful sign at present that the idea of a ruling paradigm, which is close to an imposed orthodoxy, has been more or less discarded. Alternative schools of thought coexist and this diversity is a good thing because it offers an understanding with wider dimensions.

Gerhard Hard (1973, p237) suggests that, with increasing awareness of the multitude of scientific traditions which are pursued within the framework of geography, we begin to doubt the extent to which there has been a single geographic discipline in the past. Neither a consecutive, cumulative story nor an interpretation through a dialectical approach which identifies paradigms and revolutions, provides a wholly satisfactory account of what has happened. Our perspective on the history of a discipline is always more or less influenced by the norms and outlook of the present generation. We see history from the standpoint of the present day. Whether we emphasise the continuity and gradual growth of a science or dwell on its discontinuities and revolutions, we tacitly assume a single line of progress to the present situation. Perhaps we should stress

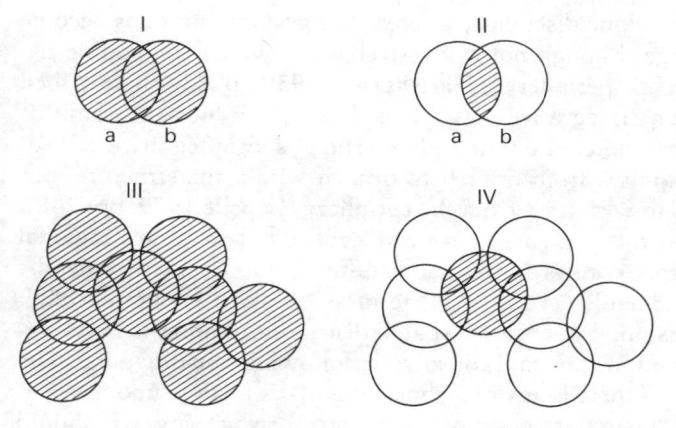

Figure 16. Different definitions of a 'geographer' (shaded). The circles symbolise closely connected research themes. a, geomorphologists; b, landscape morphologists. I–IV are explained in the text. (Adapted from Hard 1973, p235)

the heterogeneity of geography with its many-faceted and rich traditions. In the author's view, however, such a picture of the history of our discipline may also be a direct reflection of the contemporary variety of geography rather than the real story of what has happened.

Hard takes a fresh look at Fenneman's circles (figure 1, p4); are the real geographers only those who integrate all the branches of the discipline in their research? If so, there are very few real geographers. Hard uses Venn diagrams to develop his argument (figure 16). The first one is a simple and not totally unrealistic example, where the term 'real' geographer may only be applied to those who are commited to geomorphology (a), or landscape morphology (b), or both (shaded on figure 16, I). If, however, synthesis is the sole aim of geography, a 'real' geographer must study both the physical and the human morphology of the landscape in order to qualify (figure 16, II). If we consider the research themes actually being pursued by self-styled geographers the figure expands (figure 16, III) to include climatologists, geomorphologists, biogeographers, human ecologists, landscape geographers of different brands, economic geographers, locational theorists, behavioural geographers, and so on. It is too restrictive (figure 16, IV) to include only those attempting a more or less complete synthesis of natural and human factors among the 'real' geographers. Capelle (1979, p64) suggests that Fenneman's geography was an inward-looking discipline, whereas 'geography today has become outward-looking, although not aggressively so, towards compatible sister sciences on its periphery.' Hartshorne (1939, p243–5) described geography as an 'integrative' discipline, but that is not the same as regarding it as an 'integrated' discipline. There is an integrative task of transmitting impulses from branch to branch within the structure, but there is no need to withdraw from the periphery (Capelle 1979, pp67–8). We may still postulate a core or a nerve centre, but to regard regional geography as this core is no longer helpful. The core may rather be defined in Ackerman's (1963, p433) term as 'thinking geographically': 'To structure his mind in terms of spatial distributions and their correlations is a most important tool for anyone following our discipline. The more the better. If there is any really meaningful distinction among scientists, it is this mental structuring. It is one reason why we should approach analogues from other fields, as from physics, with the utmost care. The mental substrates for inspiration differ from field to field.'

This mental structuring is our trademark and an integrative effort to keep all the fast-emerging branches on our periphery within the mother science should be our common purpose. Although an integrated synthesis may be an impossible task on the research frontier, in practical life, in local planning and in school education, for example, the situation is rather different. Here the need for certain types of geographical synthesis is apparent. A synthesis is also a realistic aim where simplification rather than refinement is the objective. For this reason, basic teaching in geography at school and in universities should cover both physical and human aspects of the subject.

I have stressed human ecology and systems analysis as integrative geography, not because I think of them as geography proper, but because I regard them as important fields at the present time when man's use of the natural resources of the world may well determine his future. My personal view is that the discipline of geography can make an essential contribution to the planning of a future for mankind, and that in this context geography should base its analysis of man's activity more on ecological than on economic analogies. The reader may prefer another choice: there are a number offered. There are still, for example, many quantitative–positivist geographers who believe that we should plan our future societies with the help of models derived from economics. A growing number of geographers, on the other hand, are discarding what they term 'neo-classical economics' and seeking a strategy for a new society through the development of a radical geography based on historical–dialectical methods and Marxist philosophy. Other critics of the 'quantitative revolution' have tried to establish what has been termed humanistic geography (Ley and Samuels 1978). They have been criticised by Marxists who allege that the humanists are only trying to understand man and his lifeworld, through hermeneutic methods, but that they are not committed to changes which are both possible and necessary. Still other additions have been in the field of welfare geography and behavioural studies.

Perhaps none of these schools of thought will emerge as 'the geography of the future' but we should try to retain them all within the structure of the discipline. Overgeneralised statements about the nature of an academic discipline rarely serve a useful purpose. As to the future we may add, following Bird (1979, p124), that 'the geographer will never be the master of his fate: his very reason always progresses by leading him into the unknown and unforeseen where he learns new things.'

REFERENCES

Aase A 1970 Geografi og samfunn. Noen tendenser og problemer i dagens samfunnsgeografi *Norsk Geogr. Tidsskr.* **24** 1–21

—— 1971 Geografen — den suboptimaliserte planlegger *Samferdsel* **2** No.6

Abler P, Adams J S and Gould P 1972 *Spatial Organization: The Geographer's View of the World* (London: Prentice-Hall)

Adams J D 1968 *A Review of Behaviour and Location* LSE Graduate School Discussion Paper No.20, London

Ackerman E A 1958 *Geography as a Fundamental Research Discipline* University of Chicago, Dept of Geography Research Paper No. 53

—— 1963 Where is a research frontier? *Ann. Assoc. Am. Geogrs* **53** 429–40

Anderle, O F 1960 A plea for theoretical history *History and Theory* **1** 27–56

Anuchin V A 1973 Theory of geography, in *Directions in Geography* ed R J Chorley (London: Methuen) pp43–63

Appleton J 1975 *The Experience of Landscape* (London: Wiley)

Bagrow L 1945 The origin of Ptolemy's 'Geographia' *Geografiska Annaler* 1945 318–87

Barrows H H 1923 Geography as human ecology *Ann. Assoc. Am. Geogrs* **13** 1–14

Bartels D 1973 Between theory and metatheory, in *Directions in Geography* ed R J Chorley (London: Methuen) pp23–42

Berry B J L 1973a A paradigm for modern geography, in *Directions in Geography* ed R J Chorley (London: Methuen) pp3–21
—— 1973b *The Human Consequences of Urbanization* (London: Macmillan)
—— 1974 Review of D Harvey 'Social Justice and the City' *Antipode* 6(2) 142–5; 148
von Bertalanffy L 1968 *General Systems Theory — Foundations, Development, Applications* (New York: George Brazilier)
Bird J 1979 Methodology and philosophy, progress report, in *Progress in Human Geography* 3 117–20
Blalock H M and Blalock A 1959 Toward a clarification of systems analysis in the social sciences *Philosophy Sci.* 26 pp. 84–96
Bliss L W *et al* 1969 *Laboratory Manual for General Botany* 4th edn (New York, Holt Rhinehart and Winston).
Braithwaite R B 1953 *Scientific Explanation* (London: Cambridge University Press)
Broek J O M 1965 *Geography, its Scope and Spirit* (Colombus, Ohio: Merrill)
Broek J O M and Webb J W 1973 *A Geography of Mankind* (New York: McGraw-Hill)
Bunge W 1962 *Theoretical Geography* (2nd edn 1966) Lund Studies in Geography, Ser. C,1 (Lund, Sweden: Gleerup)
—— 1973 Ethics and logic in geography, in *Directions in Geography* ed R J Chorley (London: Methuen) pp 317–31
Burton I 1963 The quantitative revolution and theoretical geography *Canadian Geogr.* 7 151–62
Buttimer A 1978 Charism and context: the challenge of la Géographie Humaine, in *Humanistic Geography, Prospects and Problems* ed D Ley and M S Samuels (Chicago: Maaroufa Press) pp58–76

Capelle R B jr 1979 On the periphery of geography *J. Geog.* 78 64–8
Carlstein T *et al* 1978 *Timing Space and Spacing Time* (3 vols) (London: Arnold)
Chabot G 1950 Les conceptions françaises de la science géographique *Norsk Geogr. Tidsskr.* 12 309–21
Chisholm M 1975 *Human Geography: Evolution or Revolution?* (Harmondsworth: Pelican)
Chorley R J 1973a Geography as human ecology, in *Directions in Geography* ed R J Chorley (London: Methuen) pp155–69

Chorley R J (ed) 1973b *Directions in Geography* (London: Methuen)
Chorley R J and Haggett P (eds) 1965 *Frontiers in Geographical Teaching* (London: Methuen)
—— (eds) 1967 *Models in Geography* (London: Methuen)
Christaller W 1933 Die zentralen Orte in Süddeutschland (Jena) English trans. 1966 C W Baskin *Central Places in Southern Germany* (Englewood Cliffs, N J: Prentice-Hall)
—— 1962 Die Hierarchie der Städte in *Proc. IGU Symp. on Urban Geography, Lund, 1960* ed K Norborg (Lund, Sweden: Gleerup) pp3–11
—— 1968 Wie ich zur der Theorie der zentralen Orte gekommen bin *Geog. Z.* 56 88–101
Clausen H P 1968 *Hva er Historie?* (Oslo: Gyldendal)
Claval P 1980 Epistemology and the history of geographical thought *Prog. Human Geog.* 4 pp 371–84 (London, Arnold).
Coates B E, Johnston R J and Knox P L 1977 *Geography and Inequality* (London: Oxford University Press)
Coates B E and Rawstron E M 1971 *Regional Variations in Britain* (London: Batsford)

Davies W K 1972 *The Conceptual Revolution in Geography* (London: University of London Press)
Demangeon A 1905 *La Picardie et les régions voisines, Artois Cambresis, Beauvaises* (Paris: Armand Colin)
Dickinson R E 1939 Landscape and society *Scot. Geog. Mag.* 55 1–14
—— 1969 *The Makers of Modern Geography* (London: Routledge and Kegan Paul)
—— 1970 *Regional Ecology, The Study of Man's Environment* (New York)
Dray W H 1966 *Laws and Explanation in History* (London: Oxford University Press)

Entrikin N J 1976 Contemporary humanism in geography *Ann. Assoc. Am. Geogrs* 66 615–32
Eyre S R 1978 *The Real Wealth of Nations* (London: Arnold)
Eyre S R and Jones G R J 1966 *Geography as Human Ecology: Methodology by Example* (London: Arnold)

Fawcett C B 1919 *The Provinces of England* (rev. edn 1960) (London: Hutchinson)

Febvre L 1922 *La terre et l'evolution humaine*, in the series *L'Evolution de l'Humanité*, Paris English trans. 1925 *A Geographical Introduction to History* (London: Knopf)

Fenneman N M 1919 The circumference of geography *Ann. Assoc. Am. Geogrs* 9 3–11. Reprinted in *Outside Readings in Geography* ed F E Dohrs, L M Sommers, and D R Petterson 1958 (New York: Crowell) pp 2–10

Fischer E, Campbell R D and Miller E S 1969 *A Question of Place, The Development of Geographic Thought* (Arlington, Virginia: Beatty)

Fochler-Hauke G (ed) 1959 *Geographie* (Frankfurt: Das Fischer Lexikon)

Forer P 1978 A place for plastic space *Progr. Human Geog.* 2 230–67

Freeman T W 1961 *A Hundred Years of Geography* (London: Duckworth)

Freeman T W and Pinchemel P (eds) 1978 *Geographers, Bibliographical Studies* vol. 2 (London: Mansell)

Garrison W L 1959–60 Spatial structure of the economy. *Ann. Assoc. Am. Geogrs.* 49 232–9; 471–82; 50 357–73

—— 1973 Future geographies, in *Directions in Geography* ed R J Chorley (London: Methuen) pp237–49

Gould P and White R 1974 *Mental Maps* (Harmondsworth: Penguin)

Gradmann R 1931 *Süd-Deutschland* 2 vols (Stuggart: J Engelhorn)

Granö J G 1929 Reine Geographie *Acta Geographica* 2 No 2 pp. 1–202 Helsinki

Gregory D 1978 *Ideology, Science and Human Geography* (London: Hutchinson)

Guelke L 1974 An idealist alternative in human geography *Ann. Assoc. Am. Geogrs.* 64 193–202

—— 1977a The role of laws in human geography *Prog. Human Geog.* 1(3) 376–86

—— 1977b Regional geography *Professional Geogr* 29 1–7

Hägerstrand T 1953 Innovationsförloppet ur korologisk synpunkt *Medd. från Lunds Universitets Geografiska Institution. Avhandling* nr. 25. Trans. A Pred 1967 *Innovation Diffusion as a Spatial Process* (Chicago; University of Chicago Press)

—— 1973 The domain of human geography, in *Directions in Geography* ed R J Chorley (London: Methuen) pp 67–87

Haggett P 1965 *Locational Analysis in Human Geography* (London: Arnold)

—— 1972 (new edns 1975, 1979) *Geography —A Modern Synthesis* (New York: Harper and Row)

Haggett P, Cliff A D and Frey A 1977 *Locational Analysis in Human Geography* (London: Arnold)

Hannerberg D 1961, 1968 *Att Studera Kulturgeografi* (Stockholm: Scandinavian University Books)

Hard G 1973 *Die Geographie, eine wissenschaftstheoretische Einführung* (Berlin: De Gruyter)

Harris C D and Ullman E L 1945 The nature of cities *Ann. Am. Acad. Pol. Soc. Sci.* **242** 7–17

Hartshorne R 1939 The nature of geography, a critical survey of current thought in the light of the past *Ann. Assoc. Am. Geogrs.* **29** 173–658 (Lancaster, Pa)

—— 1950 The functional approach in political geography *Ann. Assoc. Am. Geogrs.* **40** 95–130

—— 1955 'Exceptionalism in Geography' re-examined *Ann. Assoc. Am. Geogrs.* **45** 205–44

—— 1959 *Perspective on the Nature of Geography* (Chicago: Rand McNally)

Harvey D 1967 Models of the evolution of spatial patterns in human geography, in *Models in Geography* ed R J Chorley and P Haggett (London: Methuen) pp549–608

—— 1969 *Explanation in Geography* (London: Arnold)

—— 1973 *Social Justice and the City* (London: Arnold)

—— 1974 Population, resources and the ideology of science *Econ. Geog.* **50** 256–77

Hempel C G 1959 The logic of functional analysis, in *Symposium on Sociological Theory* ed L Gross (Evanston)

Henriksen G 1973 *Grunnlagsproblemer og interaksjon — en metageografisk analyse·* Hovedfagsoppgave i geografi (Geografisk Institutt, Universitetet i Bergen)

Hettner A 1927 *Die Geographie, ihre Geschichte, ihr Wesen and ihre Metoden* (Breslau: Ferdinand Hirt)

Hirsch F 1976 *Social Limits to Growth* (Cambridge, Mass: Harvard University Press)

Howard E 1902 *Garden Cities of Tomorrow* (London)

von Humboldt A 1845–62 *Kosmos: Entwurf einer physischen Weltbesc-*

hreibung 5 vols (Stuttgart: Cotta) English trans. E C Otté, 1849–58 (London: H G Bohn)

Huntingdon E 1915 *Civilization and Climate* (New Haven, CT: Yale University Press)

Huxley T H 1877 *Physiography* (London: Macmillan)

James P E 1972 *All Possible Worlds: A History of Geographical Ideas* (Indianapolis: The Odyssey Press)

Johansson I 1973 Anglosaxisk vetenskapsfilosofi, in *Positivism, marxism, kritisk teori* (Stockholm: Pan/Nordsteds) pp7–67

Johnston R J 1968 Choice in classification, the subjectivity of objective methods *Ann. Assoc. Am. Geogrs.* 58 pp 575–89

—— 1978 Paradigms and revolutions or evolution *Prog. Human Geog.* 2 189–206

—— 1979 *Geography and Geographers: Anglo-American Human Geography since 1945* (London: Arnold)

Kant E 1946 Den indre omflyttingen i Estland i samband med de estniska städernas omland *Svensk Geografisk Årsbok* 22 83–124

—— 1951 Omlandsforskning och sektoranalys, in *Tätorter och Omland* ed G Enequist (Uppsala: Lundequistska Bokhandeln) pp19–49

Keltie J S 1886 Report to the Council of the Royal Geographical Society *Suppl. Papers Royal Geog. Soc.* I 450–1

Keynes J M 1936 *The General Theory of Employment, Interest and Money* (London: Macmillan)

Kirk W 1963 Problems of geography *Geography* 48 357–71

Knox P L 1975 *Social Well-Being: A Spatial Perspective* (London: Oxford University Press)

Kuhn T S 1962, 1970 *The Structure of Scientific Revolutions* (Chicago: University of Chicago Press)

Lange G 1961 Varenius über die Grundfrage der Geographie *Petermanns Geographische Mitteilungen* 105 274–83

Ley D and Samuels M 1978 *Humanistic Geography, Prospects and Problems* (Chicago: Maaroufa Press)

Liedman S E 1973 Marxism och vetenskapsteori, in *Positivism, marxism, kritisk teori* (Stockholm: Pan Nordsteds) pp68–112

Lukermann F 1958 Towards a more geographic economic geography *Professional Geogr.* 10(1) 2–10

Mackinder H J 1887 On the scope and methods of geography *Proc. Roy. Geog. Soc.* 9 141–60

Meadows D H, Meadows D L, Randers J and Behrens W W 1972 *The Limits to Growth* (London: Earth Island)

Minshull R 1970 *The Changing Nature of Geography* (London: Hutchinson)

Mishan E J 1969 *Growth: the Price we Pay* (London: Staple Press)

Montefiore A C and Williams W W 1955 Determinism and possibilism *Geographical Studies* 2 1–11

Morgan M A 1975 Values and political geography, in *Processes in Physical and Human Geography, Bristol Essays* ed R Peel *et al* (London: Heinemann) pp287–304

Morrill R L 1965 *Migration and the Growth of Urban Settlement* Lund Studies in Geography, Ser. B, 24 (Lund, Sweden: Gleerup)

—— 1974 Review of D Harvey, Social Justice and the City *Ann. Assoc. Am. Geogrs.* 64 475–7

Morrill R L and Wohlenberg E H 1971 *The Geography of Poverty in the United States* (New York: McGraw-Hill)

Myrdal G 1953 The relation between social theory and social policy *Br. J. Sociol.* 23 210–42

Newcomb R M 1979 *Planning the Past: Studies in Historical Geography* (London: Dawson/Archon)

Nordgård A 1972 *Korologiske metoder* (Oslo: Geografisk Institutt)

Odum H T 1960 Ecological potential and analogue circuits for the ecosystem *Am. Scientist* 48 1–8

Olsson G 1974 Servitude and inequality in spatial planning: ideology and methodology in conflict *Antipode* 6 (1) 16–21. Reprinted in *Radical Geography* ed R Peet 1978 (London: Methuen)

—— 1975 *Birds in Egg* Michigan Geog. Publ No. 15 (Ann Arbor)

Peel R, Chisholm M and Haggett P (eds) 1975 *Processes in Physical and Human Geography, Bristol Essays* (London: Heinemann)

Peet R (ed) 1977a *Radical Geography* (London: Methuen)

——1977b The development of radical geography in the United States, in *Radical Geography* (London: Methuen) pp 6–30

Pen J 1958 *Moderne Economie* (Utrecht: Spectrum). English trans. 1965 *Modern Economics* (Harmondsworth: Penguin)

Peschel O 1870 *Neue Probleme der vergleichenden Erdkunde als Ver-*

such einer Morphologie der Erdflöche (Leipzig: Duncker und Humblot)

Pred A 1967 *Behaviour and Location: Foundations for a Geographic and Dynamic Location Theory: Part I* (Lund: Gleerup)

—— 1969 *Behaviour and Location: Foundations for a Geographic and Dynamic Location Theory: Part II* (Lund: Gleerup)

Ratzel F 1882, *Anthropogeographie, oder Grundzüge der Anwendung der Erdkunde auf die Geschichte* (Stuttgart: Engelhorn)

—— 1891 *Anthropogeographie II: Die geographische Verbrietung des Menschen* (Stuttgart: Engelhorn)

—— 1897 *Politische Geographie* (Munich: Oldenburg)

Réclus E 1866–67 *La Terre* (Paris: Hachette)

—— 1875–94 *Nouvelle Géographie Universelle* (Paris: Hachette)

Rink F T 1802 *Emmanuel Kant's physische Geographie* (Königsberg)

Ritter C 1822–59 *Die Erdkunde, im Verhältnis zur Natur und zur Geschichte des Menschen, oder allgemeine vergleichende Geographie als sichere Grundlage des Studiums and Unterrichts in Physikalischen und historischen Wissenchaften* 19 vols. (Berlin: Reimer)

Rokkan S 1970 *Citizens, Elections, Parties* (Oslo: Universitetsforlaget)

Rostow W W 1960 *The Stages of Economic Growth* (London: Cambridge University Press)

Sack R D 1972 Geography, Geometry and Explanation *Ann. Am. Assoc. Geogrs.* **62** 61–78

—— 1974 Chorology and spatial analysis *Ann. Assoc. Am. Geogrs.* **64** 439–52

Sauer C O 1925 *The Morphology of Landscape* University of California *Publs in Geog.* **2** 19–35

—— 1963 *Land and Life* ed. J B Leighley (Berkeley)

Schaefer F 1953 Exceptionalism in geography *Ann. Assoc. Am. Geogrs.* **43** 226–49

Schilpp P A (ed) 1963 *The Philosophy of Rudolph Carnap* (La Salle, IL: Open Court)

Schmieder O 1964 Alexander von Humboldt: Persönlichkeit, wissenschaftliches Werk und Auswirkung auf die moderne Länderkunde. *Geog. Z.* **52** 81–95

Semple E C 1911 *Influences of Geographical Environment* (New York: Henry Holt)

Simmons I G 1974, 1981 *The Ecology of Natural Resources* (London: Arnold)

—— 1979 *Biogeography: Natural and Cultural* (London: Arnold)

Simpson G G 1963 Historical science: in *The Fabric of Geology* ed CC Albritton (Reading, Mass: Addison Wesley)

Skjervheim H 1974 Objectivism and the Study of Man. *Inquiry* **17** 213–39, 265–302.

Smith A 1776 *An Inquiry into the Nature and Causes of the Wealth of Nations* (London)

Smith D M 1979 *Where the Grass is Greener — Living in an Unequal World* (Harmondsworth: Penguin)

Somerville M 1848 *Physical Geography* (London)

Sømme A (ed) 1965 *Fjellbygd og feriefjell* (Oslo: Cappelen)

Spate O H K 1952 Toynbee and Huntingdon: A Study in Determinism, *Geog. J.* **118** 406–24

Stamp D 1966 Ten years on *Trans. Inst. Br. Geogrs* **40** 11–20

Stevenson W I 1978 Patrick Geddes 1854–1932, in *Geographers, Bibliographical Studes* vol.2 ed T W Freeman and P Pinchemel (London: Mansell) pp53–65

Stewart J Q 1947 Empirical mathematical rules concerning the distribution and equilibrium population *Geog. Rev.* **37** 461–85

Stoddart D R 1966 Darwin's impact on geography *Ann. Assoc. Am. Geogrs.* **56** 683–98

—— 1967 Organism and ecosystem as geographical models, in *Models in Geography* ed R J Chorley and P Haggett (London: Methuen) pp 511–48

—— 1975 'That Victorian Science' — Huxley's 'Physiography' and its impact on geography *Trans. Inst. Br. Geogrs.* **66** 17–40

Taaffe E J (ed) 1970 *Geography* (Englewood Cliffs, N J: Prentice-Hall)

Tatham G 1951 Geography in the nineteenth century, in *Geography in the Twentieth Century* ed G Taylor (London: Methuen) pp28–69

Taylor G 1951 *Geography in the Twentieth Century* (London: Methuen)

Taylor P J 1976 An interpretation of the quantification debate in British Geography *Trans. Inst. Br. Geogrs.* New Ser. **1** 129–42

Thrower N 1972 *Maps and Man* (Englewood Cliffs, N J: Prentice-Hall)

von Thünen J H 1826 *Der Isolierte Staat in Beziehung auf Landwirtschaft und Nationalökonomie* (Hamburg) English trans C M Wartenburg in 1966: *Von Thünens Isolated State* ed P Hall (Oxford: Pergamon)

Troll C 1947 Die geographische Wissenschaft in Deutschland in dem Jahren 1933 bis 1945 *Erdkunde* **1** 3–48

Tuan Yi-Fu 1971 Geography, phenomenology, and the study of human nature *Canadian Geogr* **15** 181–92
—— 1974 Space and Place: Humanistic Perspectives *Progr. Geog.* **6** 211–52
—— 1976 Humanistic Geography *Ann. Assoc. Am. Geogrs.* **66** 266–76
—— 1977 *Space and Place, the perspectives of experience* (Minneapolis: University of Minnesota Press, London: Arnold)
—— 1978 Literature and geography: implications for geographical research, in *Humanistic Geography: Prospects and Problems* ed D Ley and M S Samuels (Chicago: Maaroufa) pp194–206
—— 1980 *Landscapes of Fear* (New York: Pantheon and Oxford: Blackwells).

Ullmann E 1941 A theory of location for cities *Am. J. Sociol.* **46** 835–64

van Valkenburg S 1952 The German school of geography, in *Geography in the Twentieth Century* ed G Taylor (London: Methuen) pp 91–117
Varenius B 1650 *Geographia Generalis* (Amsterdam)
Vidal de la Blache P 1903 *Tableau de la Géographie de la France* (Paris: Hachette)
—— 1917 *La France de l'Est* (Paris: Armand Colin)
—— 1921 *Principes de la Géographie Humaine* trans. 1926 as *Principles of Human Geography* (London: Constable)

Waibel L 1933 Was verstehen wir unter Lantschaftskunde? *Geogr. Anzeiger* **34** 197–207
Wärneryd O 1977 Kulturgeografi — Samhällsgeografi, in *De Samhälls-vetenskapliga studiernas historik* (Lund) pp20–31
Warntz W 1959 *Towards a Geography of Price: a Study in Geo-Econometrics* (Philadelphia: University of Pennsylvania Press)
—— 1964 A new map of the surface of population potentials for the United States, 1960 *Geog. Rev.* **54** 170–84
Weber A 1909 *Über der Standort der Industrien* (Tübingen) trans. C Friederich (1929) as *Alfred Weber's Theory of the Location of Industries* (Chicago: Chicago University Press)
Weber M 1949 *The Methodology of Social Sciences.* English trans. of three articles published in German in 1904, 1905, 1917, ed E A Shils and H A Finch (Glencoe, IL: Free Press)
Weigt E 1957 *Die Geographie* (Braunschweig: Westermann)

White G 1973 Natural hazards research, in *Directions in Geography* ed R J Chorley (London: Methuen) pp193–216

White R and Gould P 1974 *Mental Maps* (Harmondsworth: Penguin)

Widberg J 1978 *Geografi, från naturvetenskap til samhällsvetenskap: En idéhistorisk översikt*. 3-betygsuppsats (Lund: Inst. för Kulturg. och Ekonomisk Geog.)

Whittlesey D 1929 Sequent occupance *Ann. Assoc. Am. Geogrs.* **19** 162–5

Wolpert J 1964 The decision process in spatial context *Ann. Assoc. Am. Geogrs.* **54** 337–58

Wrigley E A 1965 Changes in the philosophy of geography, in *Frontiers in Geographical Teaching* ed R J Chorley and P Haggett (London: Methuen) pp3–20

Yeates M 1968 *An Introduction to Quantitative Analysis in Economic Geography* (New York: McGraw-Hill)

AUTHOR AND PERSONALITY INDEX

SUBJECT INDEX

This index gives reference to pages where important concepts are explained or defined. It is consequently not a complete subject index.
